G. T. Wrixon, A.-M. E. Rooney, W. Palz

Renewable Energy-2000

With 14 Figures, 4 in color

Springer-Verlag
Berlin Heidelberg New York
London Paris Tokyo
Hong Kong Barcelona Budapest

Prof. Gerard T. Wrixon
National Microelectronics Research Centre,
University College, Cork, Ireland

Dr. Anne-Marie E. Rooney
National Microelectronics Research Centre
University College, Cork, Ireland

Dr. Wolfgang Palz
Commission of the European Communities,
DG XII, Brussels, Belgium

A Study prepared for EUREC -
The European Renewable Energy Centres Agency - 18.1.1993

ISBN 978-3-642-52349-6   ISBN 978-3-642-52347-2 (eBook)
DOI 10.1007/978-3-642-52347-2

© Springer-Verlag Berlin Heidelberg 1993
Softcover re print of the hardcover 1st edition 1993

Typesetting: Camera ready by author

61/3020-5 4 3 2 1 0 - Printed on acid -free paper

"Andean ingot of glory, paschal candle of the sky,
Flame at the heart of mystery, our systems burning eye."

John Montague - 'Sun Hymn'

Preface

The limited available of fossil energy carriers and the environmental impacts of energy consumtion demand mid- and long- term strategies both for the rational use of energy and for increased renewable energy utilization. Concepts of renewable energy conversion have been proposed and implemented during recent decades all over the world and their remarkable potential has been demonstrated. However, large scale implementation in the context of the requirements above has not yet taken place.

The EUREC (European Renewable Energy Centres) Agency - an association of 25 European research centres from 11 EC countries, provides a framework of leading experts in the field of renewable energies, focusing on assessment studies, joint projects in research and development and on proposing future strategies in a European context.

Under the leadership of Professor Wrixon, with the assistance and contributions of experts from the EUREC Agency, the document 'Renewable Energy - 2000' has been elaborated. This document critically evaluates the potential of renewable energies, focusing on the most relevant sources and methodologies, namely: wind energy, solar heating, cooling and daylighting, photovoltaics and biomass. It also proposes strategies for implementing components and systems to achieve economic operation in different regions of the European Community. It thereby makes use of the expert knowledge and experience concentrated in EUREC Agency member institutions.

Uncorrelated efforts in research and development certainly will supply manifold results and progress in components and systems; however, concerned actions are required if renewable energies are to play a role in counteracting limited energy resources while at the same time being environmentally compatible.

It is the feature and strength of 'Renewable Energy - 2000' that new strategies for future efforts and developments have been clearly stated. In addition, well defined concepts for promotion and implementation of renewable energies have been pointed out. The goal of the document

- to apply the highest state of the art as soon as possible in relevant projects and to serve as a guide for implementation in the energy economy under different regional conditions, has been successfully achieved by the authors. It is expected that 'Renewable Engery - 2000' will prove to be useful as a reference book for decision makers in politics and administration and will also help to co-ordinate the work of scientists and engineers to the benefit of society. It is now time for us to identify the route to travel and to realize the goals in view of the near future energy requirements of our society.

Werner H. Bloss,
Honorary President, EUREC Agency,
University of Stuttgart,
February, 1993

Acknowledgements

We wish to acknowlege all those who have assisted us in the course of this project, in particular:

A. Baubin, DGXII, CEC, Brussels
P. Bell, EOLAS, Dublin
H.J.M. Beurskens, ECN, Netherlands
W.H. Bloss, Institut fur Physikalische Elektronik, Universitat, Stuttgart
Mr. Boleau, DGXVII, CEC, Brussels
J. Bonda, EPIA, Brussels
S. Burton, ECD, London
L.Canavan, CSO, Dublin
J. Cavanagh, ETSU, UK
Dr. Colombo, ISPRA, Italy
R. Deasy, EUROSTAT, Luxembourg
B. Delmon, GEBI, Belgium
The Department of Energy, Dublin
G. Eisenbeiss, DLR, Koln
The European Documentation Centre, UCC, Ireland
E. de Oliveria Fernandes, UDP DEMEGI, Portugal
A. Frey, FhG ISE, Freiburg
A. Garrad, Garrad Hassan and Partners, London
A. Goetzberger, FhG ISE, Freiburg
G. Goss, INRA, France
J. Goulding, Energy Research Group, UCD, Ireland
J. Halliday, SERC, UK
P. Helm, WIP, Germany
W. Kleinkauf, ISET, Kassel
D. Lalas, Lambda Technical Ltd., Greece
O. Lewis, Energy Research Group, UCD, Ireland
G. Liersch, FhG ISE, Freiburg
A. Luque, Instituto de Energia Solar, Madrid
S. Mc. Carthy, Hyperion, Cork
P.H. Madsen, RISO, Denmark
D. Mayer, Ecole des Mines de Paris, France
R.P. Mertens, IMEC, Leuven
R. Montgomery, EUROSTAT, Luxembourg
B. Morgana, Conphoebus, Italy
Mr. Nacfaire, DGXVII, CEC, Brussels

X

Contents

Glossary of Symbols, Units and Abbreviations used throughout the Report

W	watts
kWh	kilowatt-hour (an average home uses approximately 4,000 kWh/year)
ton	1.016 tonnes - 1016 kilograms
toe	tonne of oil equivalent
mtoe	million tonnes of oil equivalent - 12.5 TWh
tce	tonne of coal equivalent
mtce	million tonnes of coal equivalent - 7.5 TWh
R,D&D	Research, Development and Demonstration

Prefixes:

k	kilo- (10^3)
M	mega- (10^6)
G	giga- (10^9)
T	tera- (10^{12})

Executive Summary
The Study

This Study was prepared by the NMRC, Ireland for the European Renewable Energy Centres Agency, in response to a request from the Directorate General XII (Science, Research & Development) of the Commission of the European Communities.

Its purpose is to provide a general review of current technical, economic, market penetration/commercialization status and prospects to the year 2000 for four specific Renewable Energy Sources in Europe, viz.:

(i)   wind;
(ii)   solar heating & cooling and daylighting;
(iii)   photovoltaics;
(iv)   biomass

and to assess, for each, future R,D&D needs/goals and the potential impact of Community R,D&D programmes to accelerate their technical and economic readiness.

A Summary of the Study's Key Findings

Renewable energy technologies produce marketable energy by converting natural phenomena into useful energy forms. These resources represent a massive energy potential which greatly exceeds the potential of fossil fuel resources.

It is becoming increasingly evident that Renewable Energy Technologies have a strategic role to play in the achievement of Community goals on sustained economic development and environmental protection.

A recent (1992) report to the UK Board of Trade, for instance, shows that Renewable Energy technologies could contribute between 5% and 45% of the UK's 1991 electricity supply by the year 2025 - resulting in a saving of greenhouse gas emissions equivalent to 10-100 million tonnes of carbon dioxide annually. A plausible figure is seen to be 20% of 1991 supply or about 60 TWh/year.

Total final energy consumption for Europe for 1990 was 752.77 MTOE. By the year 2000, this figure could increase by as much as 21% - posing corresponding supply/demand and environmental challenges.

Renewable energy is a domestic resource which has the potential to contribute to or provide complete security of supply.

It is virtually uninterruptible and is of infinite availability because of its wide spread of complementary technologies - thus fitting well into a policy of diversification of energy supplies.

With zero or little production of slag, ash , SOx, NOx and CO2, Renewable technologies are largely pollution-free and make zero or little contribution to the greenhouse effect with its predicted drastic concommittent climatic changes. In addition, they produce no nuclear waste and are thus consistent with environmental protection policies, building towards a better environment.

Renewable energy technologies can be used for a variety of applications that can meet practically every type of final energy demand.

Currently, renewable energy resources are - with the exception of fuel wood and large hydro electricity, virtually unutilized in Europe.

The socio-economic benefits to be accrued from increasing the utilization of Renewables in Europe are many. They include:

- positive effects to counter rural de-population, for example: job-creation through the Renewables effort;

- improved living conditions and infrastructures in the less favoured regions, for example: improved quality of life through rural electrification, etc.;

- economic benefits of avoided cost of fossil fuel imports.

Given that these energies are socio-economically desirable and will have to play an important role in Europe's future energy mix, it is important to stimulate the acceleration of market uptake. A number of specific promotional measures must be adopted, viz.:

(i) at national and Community level, policy decisions must be taken and market goals defined in order to set a general frame and express the willingness of the decision-makers to foster the development and market introduction of renewables;

(ii) a first prerequisite of market introduction is the development of well designed products already industrially and commercially available for easy utilization and integration;

(iii) regulations are needed to make the existing energy networks accesible to this new source of decentralized and renewable energy;

(iv) financial support in terms of direct subsidies for development, manufacturing and utilization of renewable energy products as well as tax deductions and cost shifting in favour of renewables through charges for conventional energies must be considered - renewables require an equitable market;

(v) norms and standards must be developed to encourage responsible and competitive market growth;

(vi) upstream of market development, prenormative R&D and scientific and technological promotion activities represent the starting point of improved performance and reliability of products and systems as well as further cost reductions.

Currently renewables account for approximately 6% of energy consumption in the EC - about half of this coming from the use of wood as a fuel source and most of the rest from hydro-electric generation. The Commission wishes this figure to increase by about 2% throughout the Community by the year 2000. Accordingly, the goal of Community action is to promote the maturity of technologies and to facilitate the uptake of renewables in Community markets.

To that end the R&D programmes of EC Directorate-General XII (Science, Research & Development) aim at improving the performance and decreasing the cost of systems and components as well as stimulating and supporting European industry in this sector.

Further, the technology-promotion programmes of Directorate-General XVII (Energy) of the Commission are aimed at bringing innovative renewable energy technologies into commercial application and at disseminating more widely (that is: within and beyond the Community) the implementation of proven technologies.

Meanwhile, the Altener programme currently coming on-line will address itself to the non-technological barriers hindering the wider implementation of renewable energy.

Impressive progress has been made within the past fifteen years in the technical development and deployment of renewable energy technologies.

Costs[*] have fallen dramatically and will continue to do so provided adequate strategies for technological development and market stimulation are adopted.

---

[*] See individual sections for reviews of economic, social and environmental costs.

Average estimated Electricity Production Costs (ECU/kWh)[**] for Renewables, range between:

- 0.05 ECU/kWh now and 0.03 ECU/kWh by the year 2000, for Wind;

- 0.6 ECU/kWh now and 0.3 ECU/kWh by the year 2000, for Photovoltaics;

- 0.1 ECU/kWh now and 0.05 ECU/kWh by the year 2000, for Biomass

compared to approximately 0.04 ECU/kWh (now) for conventional sources.

Fuels costs (from Biomass) are estimated at:

- <u>solid:</u>
  100 ECU/tce for cellulosic material (compared to 130 ECU per tonne of domestic coal);

- <u>liquid:</u>
  60 ECU/barrel of bio-alcohol from wheat;
  40 ECU/barrel of pyrolysis oil and
  refined pyrolysis oil;
  (compared to approx. 20 ECU/barrel for petrol).

Wind Energy - Status, Prospects and R,D&D Priorities

The Current Status of Wind Energy in Europe may be summarized as follows:

- 862 MW installed across Europe at present;
- peak conversion efficiencies of up to 45%;
- availabilities of over 95%;
- capacity factors of 0.15 to 0.35;
- average installed costs of 600 ECU/m^2;
- generation costs of 0.04 - 0.06 ECU/kWh.

For the Future:

- For both land-based and off-shore applications, R,D&D must continue in the areas of:

  - resource and siting;
  - strategic research on aerodynamics

[**] See relevant economic reviews for assumptions underlying these costings.

- and materials;
- turbine design and components;
- manufacturing methods (composites, etc.);
- installation, operation and maintenance;
- environmental impact (including
  noise generation, emission and propagation)
  and integration;
- energy applications and integration into
  energy systems.

leading to ...

- potential Installed Cost Reductions of 35%
  by the year 2000;

- potential Generation Cost Reductions of 25%
  by the year 2000.

Solar Heating & Cooling and Daylighting - Status, Prospects and R,D&D Priorities

The Current Status of Solar Heating & Cooling and Daylighting in Europe may be summarized as follows:

- active and passive heating systems and domestic hot
  water systems are well developed;
- active cooling systems require more work;
- additional costs for passive designs vary from zero to
  20% in extreme cases;
- costs of active systems are divided equally between
  collector, rest-of-system and installation costs;
- significant increases in deployment of solar heating &
  cooling and daylighting techologies could follow from
  R,D&D intensification for optimization and significant
  increases in promotional activities.

For the Future:

- R,D&D must continue on:

  - collection, particularly with regard to
    active systems;
  - thermal energy storage and distribution;
  - materials;
  - performance criteria, integration
    and standards;
  - design support and tools and model
    development;
  - greater market compatibility ... but also ...

- effort must be made to help these technologies
reach the marketplace:- an assessment and promotion
scheme which gives due credit for low energy design,
features and performance, must be developed and used
throughout the Community in order to heighten awareness.

Photovoltaics - Status, Prospects and R,D&D Priorities

The Current Status of Photovoltaics in Europe may be summarized as follows:

- Photovoltaic Cell Efficiencies (%):

	laboratory normal	commercial module
mono-crystalline silicon	23	15
poly-crystalline silicon	18	12
thin film:		
- amorphous silicon	13	5
- gallium arsenide	25.5	-
- CuInSe2	15	-
- cadmium telluride	15	-
- gallium arsenide (concentrator)	29	-
- TiO$_2$ organic cells	>10	-

- current module costs are 2-3 ECU/W_p;
- current Balance of System (BOS) costs (no batteries)
  are approximately 3.5 ECU/W_p;
- current overall system installation costs are
  6-8 ECU/W_p;
- current kWh costs are:
  0.4-0.6 (avoided costs - 70 ECU/m^2 or 0.5 ECU/W)
  0.5-0.7 (no avoided costs).

For the Future:

- continued R,D&D will lead to:

    - module efficiencies of over 20%;
    - lifetimes of up to or over 30 years;

- the use of abundant materials of low toxicity and a low cost production process;
- module costs of 1-1.8 ECU/W_p;
- BOS (no batteries) costs of 1.8 ECU/W_p;
- total overall system installation costs of 3.6 ECU/W_p for grid-connected systems;
- the development of concentration technology.

Biomass - Status, Prospects and R,D&D Priorities

The Current Status of Biomass in Europe may be summarized as follows:

-Sources/Production: wood; straw; sugar beet and cereal for ethanol production; rapeseed for oil production.

-Biochemical Conversion: acid hydrolysis is well established at laboratory scale and now requires demonstration; enzyme hydrolysis still requires fundamental R,D&D; ethanol production by fermentation is established commercially but further improvements could be made to reactors and separation systems, for example; particular attention should also be paid to the newly discovered means of producing ethanol from lignocellulosics by its thermal gasification to synthesis gas followed by direct anaerobic fermentation of the latter to ethanol.

-Thermochemical Conversion: the technologies for bio-oil production are evolving rapidly with improving process performance, larger yield and better quality products, justifying a substantial R&D program in this area; there is a significant need for R&D in the very promising area of gasification of biomass for use in advanced combined cycle power plants.

-Pulping: the ASAM process is a significant development here.

-Applications: work continues on optimizing existing installations to operate on new fuels and on devising and developing new applications and products as follows:

- at the crop pre-processing stage, improvements include lower cost chippers and other communition methods and cheaper and more efficient drying methods;

- at the first stage of conversion, system improvements include improved combustion and gasification technologies, gas cleaning and cooling system improvements allowing producer gas to be used in IC engines, improved anaerobic digestion systems and construction methods;

- at the final stage of conversion, ongoing developments in technologies will allow cost effective solid biomass conversion at smaller scales (<2-5MWe) than is generally possible at present - and at larger scales (about 50 MWe) for systems where large scale biomass production is viable.

For the Future:

- the principal developments in the next few years will be:

    - new sources / improved production techniques;
    - increased production of ethanol for the transport sector from high sugar-/starch- crops such as sweet sorghum;
    - use of vegetable oils such as rape-seed oil as a diesel substitute;
    - availability of various bio-oils (pyrolytic oils) in substantial amounts and the development of bio-oil stabilization and upgrading processes as well as the increased use of (refined) pyrolysis oil - for example: for turbines for peak electricity production;
    - commercialization of new lignocellulosic crops, both perennial woody short rotation crops and annual and perennial herbaceous crops;
    - more widespread harvesting of forestry residues for energy;
    - development of infrastructure for regional integration.

While the technology exists for the second and third, additional effort is required with regard to converting the fifth to electricity (to provide lower costs and higher efficiencies) and to liquid fuels (particularly in the area of hydrolysis).

It is clear from the foregoing that Renewables will continue to follow a natural growth rate, making a gradually increasing contribution to the EC energy mix over time - driven largely by technology performance improvements and being gradually manoeuved into an increasingly more cost competitive position vis-a-vis conventional sources of energy ... technologies being pushed nearer to competitive thresholds.

A conservative estimated share of Community energy consumption by the year 2000 for wind energy is about 1% of electricity demand; for active and passive solar heating & cooling and daylighting technologies: 1-2%; photovoltaics: marginal; biomass: about 3%.

Given the many political, strategic, socio-economic and environmental benefits to be accrued from the realization of the currently vastly under-exploited potential of Renewables, however, there is ample justification for accelerating this growth rate.

Growth rate acceleration may be achieved by combining political commitment and goodwill with 'Technology Push Activities' (that is: R,D&D intensification), 'Market-Pull Activities' (that is: installation subsidies, tax reliefs and premium payments) or both - and, of course, promotional activities based on well conceived, well adapted, full-scale, pilot and demonstration projects.

'Technology-push' through increased national and EC level R,D&D support and budget allocation in the areas indicated, can significantly increase the rate of market uptake of Renewables - thus impacting on the success of the Community's various Renewables Industries and European and world markets in the coming decades.

Public funding for R,D&D on Renewables in the EC in 1992 is estimated at 400 MECU. To achieve the technical goals outlined above and, consequently, an increased level of market penetration by the year 2000, this budget should be tripled.

'Market Push' means the creation of an equitable market for Renewables. Financial support in terms of direct subsidies for development, manufacturing and utilization of Renewable energy products as well as tax deductions, premium payments and cost shifting in favour of Renewables through charges for conventional energies must be considered.

Moreover, promotion activities coupled with the setting of ambitious but realistic, highly visible targets for Renewable Energy production and uptake can send important signals to industry and the general public alike.

Specific actions to accelerate market penetration of renewables in Europe include:

-    For Wind:

     A major effort could be made in the peripheral areas of the Community, particularly, the West of Ireland and Scotland.

     The current annual rate of installation of wind turbines within the Community is 300MW. It would be realistic to expect that this figure could be effectively tripled between now and the year 2000 - generating almost 2% of the Community's electricity demand of 1,100 TWh.

     A concerted effort should, therefore, be made to set up factories/wind-turbines in at least some of these areas - licensing the technology from existing European manufacturers.

     Economic studies should also be carried out for each region in order to quantify the potential benefits for disadvantaged regions of an initiative like this in terms of wind turbine production, electricity generation, avoided costs of imported fossil fuels, and so on.

-    For Solar Heating & Cooling and Daylighting:

     EC energy consumption for the heating, cooling and lighting of buildings is substantial (approximately 50% of all primary energy).

     The Community's building stock is replaced at a rate of about 2% per annum. This provides an opportunity for the introduction of legislation regarding energy requirements of new buildings - where building would not be permitted unless basic energy criteria were met. This could be preceeded by a period of optional compliance.

     Energy auditing could also be introduced for existing buildings - on a voluntary basis at first - with guidelines on energy consumption/m^2/year.

     Both of these activities could then be used to gradually frame broader legislation.

     If EC Member States were to adopt the modest and, certainly, realistic goal of introducing mandatory energy criteria for all new buildings built after the year 2000 so that energy consumption could be reduced by half the current level, cumulative energy savings of up to 1% per annum could be achieved.

12

- __For Photovoltaics:__

If PV R,D&D efforts/budgets were intensified by a factor of 3, there could be, by the year 2000, substantial increases in the use of PV:

> -on roofs - solar tiles: at least 10% of all new domestic buildings each year, in each Member State;

> -in building cladding: at least 10% of all new offices and industrial buildings each year, in each Member State;

> -in centralized power stations: at least five new PV power stations of at least 5 MWp in Southern Europe, by the year 2000.

Then, higher production volumes would in turn mean further cost reduction and even greater deployment.

- __For Biomass:__

A modest and certainly an attainable goal would be the adoption of actions across the Community, following those of France, aimed at biomass substitution of at least 5% of imported hydrocarbons by the year 2000.

With regard to electricity generation from biomass, a realistic target would be that by the year 2000, at least 1% of each Member State's electricity requirements be generated from biomass sources.

In this regard, there is an urgent need for a number of good biomass demonstration plants across Europe - to demonstrate potentials and accelerate movement along the learning curve of how to implement such plants efficiently and well - a realistic implementation target would be at least twenty small (1-5 MW - for islands and developing countries) and five large (50 MW maximum - to act as a prototypes for local generating stations in Europe) demonstration plants by the year 2000.

Finally, Renewables must be viewed as 'alternative' only to traditional fossil fuel sources. They are, in fact, complementary to each other - and can be used effectively alone or in combinations of two or more (wind and biomass, for example). Thus all Renewable options should be pursued in tandem.

1 Introduction

1.1 Background

1.1.1 Current policy, programmes and targets for renewable energy within the Community

Policy and Programmes

The Community's main rationale for developing renewables is to increase the security of energy supply for its Member States and to contribute to a better environment.

To that end the R&D programmes of EC Directorate-General XII (Science, Research & Development) aim at improving the performance and decreasing the cost of systems and components as well as stimulating and supporting European industry in this sector.

Further, the technology-promotion programmes of Directorate-General XVII (Energy) of the Commission are aimed at bringing innovative renewable energy technologies into commercial application and at disseminating more widely (that is: within and beyond the Community) the implementation of proven technologies.

Meanwhile, the Altener programme currently coming on-line will address itself to the non-technological barriers hindering the wider implementation of renewable energy.

Between 1979 and 1989 the Energy Directorate-General (DGXVII) of the Commission operated a Community programme promoting demonstration and industrial-pilot projects in the energy field. Under this programme financial support totalling 285 MECU was offered to 934 projects in the renewable energy field. Given that this financial support was up to a maximum level of 40% of eligible project costs, these projects represent investments of at least 700 MECU.

The energy demonstration programme finished at the end of 1989 - although projects financed through it will still be coming to maturity for a number of years to come.

In 1990 a new programme was launched:- Thermie for the promotion of European energy technologies (approximately 25% of Thermie's four-and-a-half year 700 MECU budget being allocated to renewable energy sources).

Since 1975, Directorate-General XII (Science, R & D) of the Commission has been implementing Community programmes for renewable energies and has given financial support of approximately 250 MECU as shared-cost contracts.

It's recently concluded programme - JOULE I ran from 1989 to March of this year and included 47 MECU for renewable energies.

In September 1991, within the third framework programme, the Council of Ministers adopted a new non-nuclear energy programme - JOULE II covering the period until the end of 1994. 57 MECU of this has been earmarked for renewable energy development.

Targets for Renewables

Currently renewables account for approximately 6% of energy consumption in the Community - about half of this coming from the use of wood as a fuel source and most of the rest from hydro-electric generation.

The Commission wishes this figure to increase by about 2% throughout the Community by the year 2000. Accordingly, the goal of Community action is to promote the maturity of technologies and to facilitate the uptake of renewables in Community markets.

1.1.2 The case for renewables

Most of the Renewable Energy Technologies which have emerged in recent years (photovoltaics, solar energy applications in buildings, biomass, wind) are vastly under-utilized at present in Europe's energy markets (see Section 1.1.1 -'Targets for Renewables').

The fact that these appropriate and convenient energy sources have been neglected until now seems surprising. The reasons are:

(a) most of the recent past has been a period of energy abundance with cheap oil available until the mid-70s and again in the mid-80s together with the anticipation of prospects for nuclear power - though the early hopes for this latter technology could not be confirmed in practice;

(b) most of the technology required to develop cost-competitive market products from solar, wind and biomass sources is relatively recent:- ten years ago appropriate technologies for wind and biomass did not exist, solar heating was too expensive and energy from biomass was not a viable option as priority had to be given to food production at that time.

With reference to the first reason given, the fact that Europe must now face the prospect of more complex and more expensive energy supply and utilization schemes (given possible supply problems and environmental concerns) means that it is now time to develop and implement new energy strategies (see Appendices 2, 3 and 4).

In future energy scenarios conservation and rational use of energy will be most important - but on the supply side, renewable energy resources must assume an increasing role as quickly as possible. In 1990, total final energy consumption was 752.77 MTOE. By the year 2000, this fugure could increase by up to 21% - that is: to 907.83 MTOE (see Appendix 5).

Renewable Energy is a domestic resource which has the potential to contribute to or provide complete security of supply. It is virtually uninterruptible and is of infinite availability because of its wide spread of technologies - thus fitting well into a policy of diversification of energy supplies. Its technologies are largely pollution-free and thus consistent with environmental protection policies.

With reference to the second reason given above, conventional wisdom in the energy field has it that most renewable energies are too expensive, require too much land for installations, are not really pollution-free, etc., etc..

Palz (1991) addresses some of these issues viz.:

(i) Low energy density:

Because renewables are not centrally available, except, for example, large hydro and tidal plants, it is claimed that renewables cannot benefit from economies of scale. On the other hand, it should be recalled that energy utilization is decentralized. In this respect, conventional central power plants are not really 'user-friendly' and require very large and costly distribution networks.

Renewable energies can do without such networks and the huge cost associated with them. However, renewables are flexible enough in use ot be integrated into existing networks if so wished.

(ii) Intermittency:

A number of renewable energies such as solar heating, photovoltaics or wind energy are intermittent. Hence, on a small scale, energy storage systems must be provided if continuous power supply is required. For large-scale utilization, however, the combination of various complementary forms of renewable energies - some of which can be made continuously available (biofuels, hydropower, etc.) reduces or completely eliminates the storage problem.

In the case of electricity networks where continuous supply is essential, a detailed study undertaken from 1988-1990 by European and US renewables and utility experts confirmed that event the most intermittent sources such as wind and photovoltaics could be used in industrialised countries in very large amounts (up to 10-15% of current power capacity) without particular implementation problems.

(iii) The need for large land areas for installation:

All renewable energies require collecting areas for solar radiation or wind but these areas do not exceed those currently taken up for the installation of conventional energy systems (mining, transport, fuel storage, conversion, distribution, disposal areas, etc.). There is one exception, however:- biomass, where the best collection efficiencies of solar energy are just 2-5%.

More generally, though, most solar energy collectors for heating or photovoltaics can be, unlike conventional energy systems, easily integrated into existing building structures and in particular the roofs of buildings so that no land requirement is involved at all. Photovoltaics could also be integrated into other existing structures (along highways, etc.) which otherwise would not be used.

Dual use of land is also a possibility in the case of wind energy:- there are many instances of wind turbines being installed on land also employed for agriculture, cattle raising and so on.

(iv) High cost:

A state of cost-competitiveness with conventional energy sources has now been reached by wind power, biomass and solar energy applications in buildings - though, at present, photovoltaics is cost-competitive for remote applications only.

Indeed, for large-scale implementation, renewable energy systems offer additional cost advantages when compared to their conventional counterparts:- renewables, because they are small and decentralized in nature, involve easier and faster implementation. Moreover, they are installed close to the consumer and hence can be more easily adapted to the users' changing needs. Renewables enable short-term realistic planning.

Construction times for conventional power plants are in the order of ten years and it is difficult to readjust planning once construction has starte. In the past - as a consequence of erroneous energy consumption projections, a lot of overcapacity of conventional power plants has been built. The large excess cost incurred in the process could have b een avoided if the short-term planning which is possible with renewables had been available.

The external cost of energy production must also be counted in favour of renewables:- such costs are transparent and totally quantifiable in the case of renewables.

1.1.3 The Present Study

The present Study represents an evaluation of the present and projected year-2000 performance status of the four renewable energy technologies:

    (i)   wind;
    (ii)   solar heating & cooling and daylighting;
    (iii)   photovoltaics;
    (iv)   biomass

and their potential contribution to the satisfaction of the Community's energy requirements over the next decade.

These four sources were chosen as, in the medium- to short- term, they represent the most technologically feasible - as well as the most commercially promising of the range of renewable energy options on offer (see Appendix 1).

The Study was prepared by the NMRC, Ireland for the European Renewable Energy Centres Agency, in response to a request from the Directorate General XII (Science, Research & Development) of the Commission of the European Communities.

The Study's assessments are based on data from the following sources:

- EUROSTAT Energy Publications;
- IAEA, ETSU and other similar bodies;
- Conference Proceedings;
- European Directory of Renewable Energy Suppliers and Services;
- EC Energy Newsletters;
- other sources as listed in the 'Bibliography/References' and 'Acknowledgements' sections of this Report.

Technological factors such as resource access, conversion efficiency, lifetime/reliability, market compatibility and manufacturability are reviewed and the Research, Development and Demonstration (RD&D) actions that would remove key technological constraints on implementing those energy resources that are the object of the Study, are identified.

The Study also identifies a number of other constraints - largely institutional (regulatory, financial, infrastructural and perceptual), which can be reduced or removed with appropriate actions.

1.2 Aims and Objectives of the Present Study

Following from section 1.1.3, the purpose of the present study is to carry out a general review of current technical, economic and market penetration/commercialization status and prospects to the year 2000 for four specific Renewable Energy Sources in Europe, viz.:

- wind;
- solar heating & cooling and daylighting;
- photovoltaics;
- biomass ...

and to assess, for each, future R,D&D needs/goals and the potential impact of Community R,D&D programmes to accelerate their technical and economic readiness.

1.3   Outline of the Report

The Report's introductory section outlines the background to the Study in terms of current policy, programmes and targets for renewables within the Community and the case for renewables generally.

Within the main body of the Report, a full section is given to each of the four Renewable Energies under review - each giving a description of the technology, its current status and contribution potentials.

The key findings of the Study overall are then drawn together to present a series of Conclusions on the prospects for the development of renewable energy in Europe to the year 2000.

R,D&D targets - and strategies for their achievement are outlined, and the possible role of the EUREC Agency in a global R&D strategy for Renewables in Europe is addressed.

Fig. 1. Wind Resources at 50 metres above ground level for five topographic conditions

---

[1] In addition to land based wind energy, there is an enormous offshore resource. The offshore potential has the advantage of higher wind speeds but the disadvantage of more difficult access.

2 Wind Energy

2.1 Technical Review of Wind Energy

2.1.1 Current Status

Resources:

Figure 1 gives an overview of Europe's wind energy potential. [1]

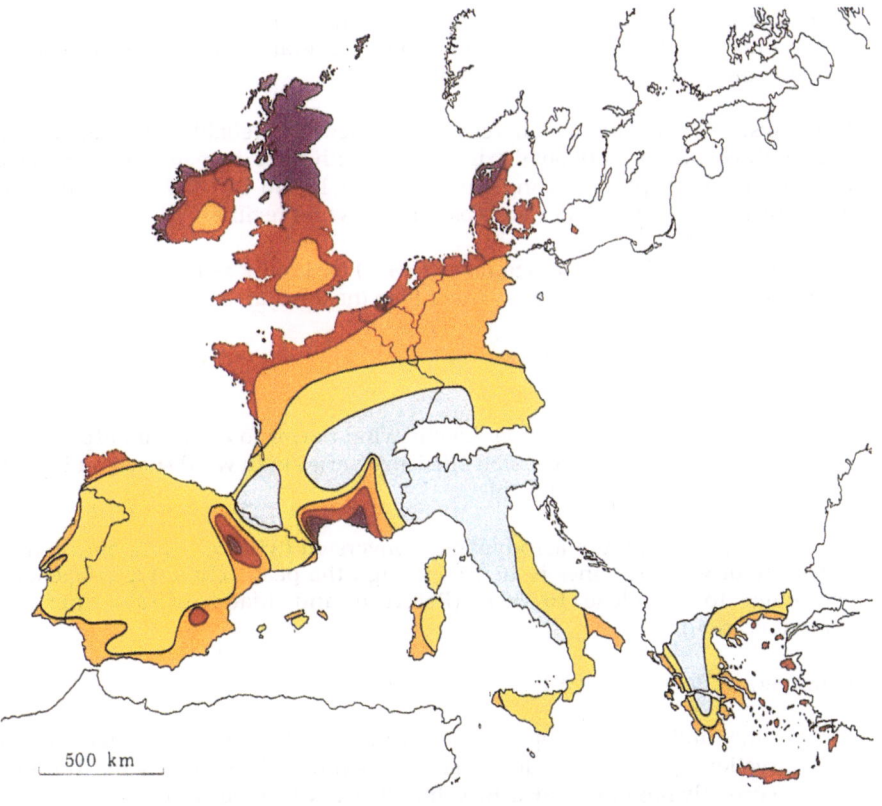

500 km

	Sheltered terrain²		Open plain³		At a sea coast⁴		Open sea⁵		Hills and ridges⁶	
Wind resources¹ at 50 metres above ground level for five different topographic conditions										
	m s⁻¹	Wm⁻²	m s⁻¹	Wm⁻²	m s⁻¹	Wm⁻²	m s⁻¹	Wm⁻²	m s⁻¹	Wm⁻²
	> 6.0	> 250	> 7.5	> 500	> 8.5	> 700	> 9.0	> 800	> 11.5	> 1800
	5.0-6.0	150-250	6.5-7.5	300-500	7.0-8.5	400-700	8.0-9.0	600-800	10.0-11.5	1200-1800
	4.5-5.0	100-150	5.5-6.5	200-300	6.0-7.0	250-400	7.0-8.0	400-600	8.5-10.0	700-1200
	3.5-4.5	50-100	4.5-5.5	100-200	5.0-6.0	150-250	5.5-7.0	200-400	7.0-8.5	400-700
	< 3.5	< 50	< 4.5	< 100	< 5.0	< 150	< 5.5	< 200	< 7.0	< 400

1. The resources refer to the power present in the wind. A wind turbine can utilize between 20 and 30% of the available resource. The resources are calculated for an air density of 1.23 kg m⁻³, corresponding to standard sea level pressure and a temperature of 15°C. Air density decreases with height but up to 1000 m a.s.l. the resulting reduction of the power densities is less than 10%, see Table B.1 in Appendix B.
2. Urban districts, forest and farm land with many windbreaks (roughness class 3).
3. Open landscapes with few windbreaks (roughness class 1). In general, the most favourable inland sites on level land are found here.
4. The classes pertain to a straight coastline, a uniform wind rose and a land surface with few windbreaks (roughness class 1). Resources will be higher, and closer to open sea values, if winds from the sea occur more frequently, i.e. the wind rose is not uniform and/or the land protrudes into the sea. Conversely, resources will generally be smaller, and closer to land values, if winds from land occur more frequently.
5. More than 10 km offshore (roughness class 0).
6. The classes correspond to 50% overspeeding and were calculated for a site on the summit of a single axisymmetric hill with a height of 400 metres and a base diameter of 4 km. The overspeeding depends on the height, length and specific setting of the hill.

Wind technology in Europe has gained considerable maturity over the last decade. The industry has gone through several shake-out phases and has gained real strength through market developments in California as well as its Member States - most notably Denmark. Countries like Germany, the UK, the Netherlands, Spain and Greece are quickly catching up.

Today the total installed wind capacity across the Community's twelve Member States stands at approximately 800 MW, generating some 1400 GWh per year.

Several of the Member States have committed themselves to ambitious expansion programmes in line with the Community's overall stated goal of 4000MW installed capacity by the turn of the century.

Wind systems available commercially at present are reliable intermediate size two- or three- bladed horizontal axis turbines, with rotor diameters in the range 20 to 39m and with power ratings in the range 150 to 500 kW. They are cost-competitive if operated under the most suitable wind regime.

Wind turbines in the MW range (over 1 MW) exist only as experimental units - their costs are three to four times greater than the medium size machines.

Efficiencies:

Peak conversion efficiencies (from kinetic wind power to electrical output) of up to 45% are achieved but conversion efficiency varies with wind speed and typical values are generally lower than this.

The theoretical maximum aerodynamic conversion (wind to mechanical power) efficiency of wind turbines is 16/27 - though the peak achievable efficiency is estimated to be close to 50% (Petersen and Madsen, 1992, personal communication).

However, as Beurskens (1991) observes:

> - the fundamental properties of even the most efficient of modern aerofoil sections (used for blades of large and intermediate size wind turbines) currently limit the peak achievable efficiency to about 45%;

> - in practice, the need to economize on blade costs tends to lead to the construction of slender bladed, fast running wind turbines with peak efficiencies a little (say, 3 or 4%) below the optimum.;

> - the average year round efficiency of most turbines is about half this figure - principally because of the need to shut down in high winds and the need to limit the power once the rated level is reached;

> - reductions in efficiencies are also caused by generator and gearbox losses and by the fact that machines do not always operate in their optimum working point.

More important, however, is the ratio of energy produced over a year to kinetic energy in the wind. In the 300-400 kW size range, this latter ratio is slightly above 30%. This average energy efficiency depends on the variability of the wind resource as well as the technical performance and design of the machine. Present machines have average efficiencies close to the theoretically achievable.

Capacity Factor:

Today's capacity factors vary from 0.15 to 0.35 (USA). The capacity factor can be chosen by varying the installed rated power per m^2 swept rotor area for given average wind speeds. A high capacity factor implies a somewhat lower energy production.

Machine Characteristics, Wind Speed, Availability, Output and Performance:

It is generally thought that intermediate size machines are approaching the limit of their technical performance and there is little scope for performance improvement.

It should be noted, however, that today's grid-connected wind-turbine generators vary considerably in size ... from machines able to produce less than 1kW of power to large wind turbines with the capacity to generate up to 3MW.

Energy output across all machines is, of course, very sensitive to the available wind resource (it is proportional to the cube of the wind speed).

The average specific annual outputs (kWh/m^2) over the last four years for the ten best wind turbines in Europe are over 1000.

Detailed Performance Records for Machines in Europe are available through EUROWIN - the European Wind Turbine Database (see Schmid and Klein, 1991).

While early systems suffered from technical problems such as blade failure, control system failure, etc., and availability was as low as 50%, such problems have largely been solved.

In the USA and Europe, availabilities of over 95% have been recorded over several years.

A wind power plant which is grid-connected can be treated in the same manner as any other power plant vis-a-vis calculating the chance that the plant, or a part of the plant, cannot supply the expected load and consequently the other plants in the network must take over.

Because of the variability of the wind, an installed MW of wind power has a lower statistical availability than a MW of a conventional plant. Any MW of wind added to the system, however, increases the reliability of the system as a whole and thence, the total system's capacity credit.

The capacity credit of a wind turbine or any added unit is equal to the capacity factor multiplied by the rated power, if the total added power does not exceed, say, 10-15% of the total power installed.

Today, availability for grid-connected wind farms has reached 95% :- hence poor availability is no longer a constraint on the widespread implementation of this technology.

Fig. 2. The Danish one-row wind farm at Kappel in Jutland. The farm consists of 24,400 kW wind turbines

2.1.2 Development Prospects

Table 1 summarizes the principal current technological constraints and opportunities for Wind Energy, based on the 1990 U.S. Interlaboratory White Paper SERI/TP-260-3674 DE90000322 on the Potential of Renewable Energy.

Table 1. The Principal Current Technological Constraints and Opportunities for Wind Energy.

Issue	Constraint	Opportunity
Resource Access	Need to get more output from each site	R,D&D on large arrays, complex terrain and wake effects
Conversion Efficiency	Need to get more output from each machine	R,D&D on airfoils, controls and advanced generators
Reliability/ Lifetime	Need to improve lifetime	R,D&D on structural dynamics, fatigue, manufacturing, transmissionless designs and load estimation for various conditions
Market Compatibility	Integration issues with utilities	R,D&D on utility integration issues (including isolated grids) and on cost reduction

The issues identified in Table 1 are discussed in detail in Section 6.

2.2 Economic Review of Wind Energy

2.2.1 Cost Determinants

The generation costs of wind energy are determined by:

- the total initial investment cost;
- economic parameters such as interest rate, amortization period, etc.;
- system efficiency of the wind turbine system;
- wind speed;
- overall expected annual average power output;
- technical availability;
- O&M costs;
- lifetime.

(i) Total Initial Investment Cost

Investment costs include wind turbine costs ex-works, foundation costs, transport costs and erection/installation and infrastructure costs which include the cost of grid connection.

Present machine costs are 300-600 ECU per m^2.

The cost of foundations depends on conditions at the site (such as stability and load-bearing capacity of the subsoil) and may vary considerably. The spread increases with the diameter which means that the security factors differ considerably for different foundations.

Infrastructure costs will add 30%, giving an average installed cost of 600 ECU per m^2.

The infrastructure costs of remote sites with high wind speeds will, of course, be higher - giving average total costs of 800 ECU per m2, while the infrastructure costs of decentralised sites will be lower since there are likely to be local grid connections and roads will not be needed (these sites will not, however, have such favourable wind regimes and unit energy costs are therefore likely to be higher).

In 1990, the total investment costs of the turbine were approximately 30% above the price of turbine ex-works. Thus, approximately 23% of total investment costs were due to foundation, installation/grid-connection, etc..

(ii) Economic Parameters

These include interest rates, amortization period, etc. and are generally based on expectations of an economic lifetime of 20 years (although the proven technical lifetime of the best machines is at the moment about 10 years) and an interest rate of 0.05% per annum.

(iii) System Efficiency of the Wind Turbine System

System efficiency estimates are based on:

- average power outputs per m^2 swept rotor area (W/m^2) as a function of the wind speed;

- air density (kg/m^3);

- energy pattern factors (the ratio between the actual energy content of the wind for a given time period and the energy of the wind calculated on the basis of the constant wind speed and correspond to figures presented in section 2.1 of this Report.

(iv) Wind Speed

As wind energy potential is proportional to the annual average wind speed cubed $(m/s)^3$, it is vital to choose sites with the highest possible wind speed.

In 'average' European cases, the average good wind speed at 10m height is 5.5 m/s. The annual energy output of a typical wind farm consisting of typical 300kW/25m machines based on this wind speed and average performance would be about 700-800 kWh/m^2 swept rotor area.

(v) Overall Expected Annual Average Power Output

Overall expected annual average power output depends on the power characteristics of the wind turbine system and on wind speed.

(vi) Technical Availability

Experience in the USA has shown that the best machines reach availability levels of 95% after five years.

(vii) O&M Costs

Annual O&M costs are often taken as a percentage of the initial investment cost.

Experience from wind farms in Denmark and the USA indicate an average O&M cost of 2.5% of capital cost per year, or 0.5-1 ECU/kWh.

Electricity companies generaly determine the O&M cost as a fraction of the generation cost per kWh. Thus, for windier sites unit O&M costs will be lower due to the higher electrical output.

Overall quoted O&M costs based on practical experience vary from ECU 0.0004/kWh for mature European wind turbines to very high figures for failed projects.

(viii) Lifetime

With regard to 'proven technical lifetime', several modern wind turbines have been in operation for 10 years without serious problems.An economic lifetime of 20 years is generally used for the analysis of generation costs, however.

2.2.2 Overall Generation Costs

There is considerable variance in average generation costs. At present, for instance, the accepted socio-economic cost in Denmark, is 0.05-0.065 ECU/kWh. For the best machines, generation costs lower than 0.05 ECU/kWh are possible.

Overall near future Generation Costs (ECU/kWh) have been estimated as follows:

(i) circa 0.05 for state of the art in Denmark,
the Netherlands, Spain and North Germany;

(ii) circa 0.06 for 'low wind' countries such as
France and Italy;

(iii) circa 0.04 for 'high wind' countries such as
Ireland, Scotland and Western Wales and England.

Thus, while wind energy can represent a viable alternative to coal-fired or nuclear power plants at sites where advanced technology and a good local wind energy potential are combined, the cost of generating electricity from wind (that is: the economic cost of wind energy) is at present, approximately 40% higher than it would be with traditional fossil energy sources.

In the case of a total cost (includes government and external cost) comparison, wind energy is already one of the cheapest energy resources available to Europe.

Finally, it is to be noted that implementation and O&M costs can be up to 55% greater for offshore wind farms than for land based plants - but that the overall economic viability is held to be comparable to that of land based wind farms (see Jensen, 1990).

Future cost reductions in wind systems are likely to result from:

- falling machine costs due to series production;
- larger machines on taller tower with access to up to 30% more energy, increasing the load factor and reducing the installed cost per MWh/year;
- improvements in the performance of larger machines resulting from advanced generator technology and transmissionless design concepts;
- optimization.

Reductions in infrastructure costs are unlikely as the civil works are standard engineering practice.

By the year 2000, there should be an overall wind generator cost reduction of up to 25%. Further, the development of very light, flexible machines with just half the tower head weight of existing machines could also reduce factory gate prices. The potential cost reduction factor is, however, controversial - estimates ranging from 15% up to 40%.

2.3 Environmental Impacts and Public Acceptability of Wind Energy in the European Context

The principal environmental impact / public acceptability issues relating to wind energy in Europe are:

- energy costs of materials;
- land requirements;
- visual impacts;
- noise;
- electromagnetic interference (television and radar, for example);
- safety;
- impact on wildlife and natural habitat.

(i) Energy Costs of Materials

Schmid and Klein (1991) have shown that energy input for the principal components used in wind energy generation, is generally:

- largest for the tower (steel/concrete),

- followed by the nacelle (steel/copper and including hub and subcomponents),

- then rotor blades (grp) and

- foundations (concrete).

The 'energy payback time' of wind turbines varies from a few months to one or two years at most (one Danish study concluded 100 days for a 100 kW machine).

The findings of a study by the University of Groningen (cited in Beurskens, 1991), indicate that large wind turbines have smaller energy payback times than small ones (the energy payback time of a 10 kW turbine was found to range from 0.5 to 1.5 years whereas the equivalent figure for a 500 kW machine was just 2 - 3 months).

(ii) Land Requirements

Regarding land requirements, it should be noted that, of the total area covered by wind machines, about 0.2% is attributed to the turbines themselves - typically: (70,000 m^2 per MW x 0.2%).

Although some land is required for access roads, utility buildings, etc., the area available for additional usages (agriculture, for example) is significantly larger (about 99%).

(iii)   Visual Impacts

Regarding visual impacts, recent study by the University of Leiden (cited by Beurskens, 1991) found that:

-    public appreciation of a landscape incorporating wind turbines depends less on the size of the turbines and more on the numbers - and is less, the more wind turbines are installed .. thus the study concluded that it is best to install large (1 MW size) machines;

-    turbine installation in open or half-open landscapes results in a much greater reduction of the appreciation that with installations in industrial areas - thus wind turbines appear to be more acceptable when sited in a 'modern man-made landscape' than in a more 'natural' location;

-    preservation of the coherence and clarity of the landscape is important ... respondents in the study expressed a preference for turbines in lines - particularly along roads and canals, rather than cluster or park formations.

Westra and Arkesteijn (1992) have further indicated that three-bladed turbines are more favourably accepted than two-bladed turbines and that well-designed turbines which are kept turning can have a positive influence.

(iv)   Noise

A typical modern 300 kW windmill operating at a wind speed of 8 m/s produces a noise level of about 45 dBA at a distance of about 200 metres. Noise levels for a windfarm of 30 such machines the corresponding figure would be 45 dBA at 500 metres from the nearest machine.

The nuisance level of the technology is determined by:

(i)   the acoustic emission of the turbine(s) and

(ii)   locally acceptable immission levels.

When emission and immission levels are known, minimum nuisance distances can be calculated and wind turbines sited accordingly.

Regulations governing noise vary considerably across the Community - and may not, of course, reflect noise levels acceptable to an individual.

The approach to noise control favoured on mainland Europe appears to be that of setting fixed noise limits which cannot be exceeded, while the preferred approach in the United Kingdom, for example (as embodied in the British Standard BS 4142), relates permitted noise limits to existing background noise levels at the site.

The difficulty in defining a universal noise criterion for wind energy technologies is clearly demonstrated by the wide variety of noise criteria that have already been agreed between developers and planning authorities at various sites throughout the Community.

(v)   Electromagnetic Interference

Wind turbines can interfere with television reception and public service links.

In general, wind turbines should not be sited in the line of sight of a microwave link - in this way problems with public service transmissions can be avoided.

Television broadcasts are, of necessity, multi-directional and it is therefore possible that television reception will be adversely affected for a number of dwellings at some sites.

In most cases, such problems can be addressed by cheap and simple technical solutions such as retuning to a different transmitter or, if necessary, fitting a more directionally discriminating aerial.

(vi)   Safety

The safety record of wind energy technology is generally good.

A number of serious and fatal accidents have occurred, however.

The European Wind Energy Association's publication 'Wind Energy in Europe', suggests that such accidents tend to result from poor management or non-observance of safety regulations, rather than from technical faults... though it is also argued that in most cases, accidents result from inadequate quality control during design and production - as well as during installation and maintenance (Beurskens, 1992, personal communication).

Modern wind turbines are equipped with monitoring systems which are intended to give early warning of potential failures of most machine components. Though generally employed for operational reasons - to reduce maintenance costs, such monitoring systems could also be used to monitor machines for potential safety hazards.

Moreover, wind turbines are increasingly being certified by independent classification societies - though certification systems have not, as yet, been harmonized.

In 1987, an EC project was initiated to draft a conceptual wind energy safety standard.

Meanwhile, a number of design and construction standards for wind turbines have also been developed by individual Member States within the Community and the the International Electrical Committee (IEC) is developing an official international safety standard on safety.

(vii)   Impact on Wildlife and Natural Habitat

The period of construction for a windfarm is typically less than a year. The process disrupts just a small proportion of the land designated for the plant and once construction is completed, the site can be returned to its former condition with the exception of a small area.

It is, however, important that windfarm developments are carefully planned and monitored to prevent unnecessary intrusion or disruption.

The impact of windfarms on bird life is an area of particular concern:- injuries or death can result from collisions against tower or blades and breeding, resting or feeding birds can be disturbed in the vicinity of the turbine.

A study by Denmark's Ornithological Association has, however, shown that birds do get used to the machines and learn, as a matter of course, to fly around them.

All of the issues discussed above can be dealt with by careful designing and planning.

Public acceptance of the concept of wind energy as a pollution-free technology is widespread - attitudes can be different, however, when it comes to a particular development.

In cases where the public is insufficiently informed, opposition to new projects is to be expected. A new, highly visible and little understood technology is naturally viewed with a degree of unease and suspicion.

It is, in fact, not generally known that wind power is technically mature, safe and economically viable.

The European Wind Energy Association's publication 'Wind Energy in Europe' affirms that location, information dissemination, the manner in which a project is announced and the way in which people are allowed to participate in the decision making process are all important.

A study carried out by the University of Amsterdam (again, cited in Beurskens, 1991) shows that public acceptance before the actual realization of projects is significantly greater in cases where the public is informed of the plans in advance of press releases and the appearance of newspaper articles. Interestingly, it is also commonly found that public acceptance increases sharply after turbine erection - apparently, as unfounded fears of the unknown are allayed.

2.4   The Commercialization of Wind Energy in Europe

The current installed capacity in the Community is about 862 MW. Specific market development programmes in Denmark, Germany, the United Kingdom and the Netherlands have resulted in the greater part of Europe's wind power capacity being concentrated in these countries - the majority (450 MW) in Denmark, 120 MW in Germany and 100 MW in both the Netherlands and the United Kingdom (see Figure 3) .

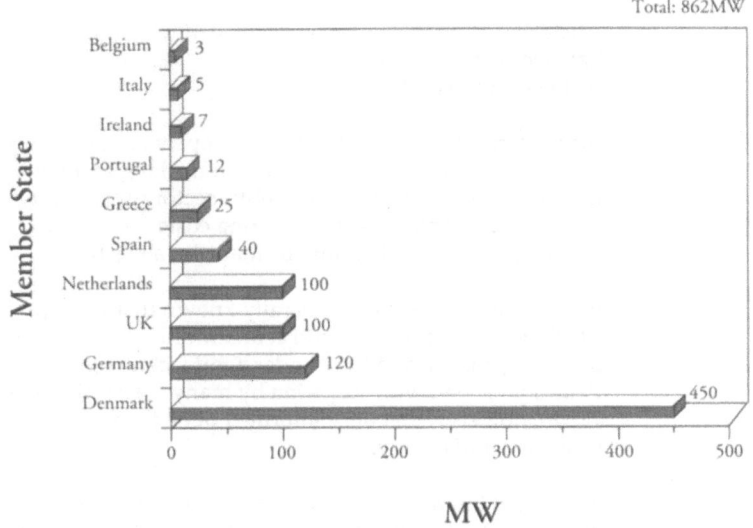

Fig. 3. Wind energy - EC installed capacity (MW) 1992

It is expected that around 4000 MW of capacity will be installed in the Community by the year 2000 - thus bringing the estimated share of Community energy consumption in 2000 to about 1% of electricity demand (an increase in R,D&D and market stimulation activities as outlined in Section 6 of this Report would, however, serve to hasten and increase the deployment of wind energy across the Community).

There are three distinct markets for wind power:

-       grid-connected in the industrialised world;

-       grid-connected in the developing world;

-       stand-alone systems.

The European wind energy industry leads the world. Given buoyant home markets, it would be in a strong position to exploit substantial export markets.

It is, however, *within* the EC that the major market developments are envisaged for foreseeable future. Indeed, in the short term, the EC represents the largest potential market for wind power in the world.

The principal barriers to a more widespread deployment of wind energy in Europe, have been identified by the European Wind Energy Association, in its publication 'Wind Energy in Europe' as:

- a distorted market - the pricing policy of which demands that electricity be provided at the lowest cash cost and not the lowest total cost;

- financial injustice - wind power developers are usually private sector entrepreneurs to whom the long and low-interest-rate loans available to public sector projects are not available ... comparison of generating costs for various technologies cannot therefore be made on a fair basis;

- public mistrust - it is not generally known that wind power is technically mature, safe and economically viable ... and so this highly visible and little understood technology is naturally viewed with suspicion by many ... experience has shown, however, that once informed of all the facts, people welcome wind energy;

- entrenched interests - wind power developers attempting to gain access to the Community's electricity grid networks have often been met by conservative dogma, lack of understanding and lack of vision;

- obstructions to free trade - existing design and construction standards for wind turbines established individually by EC Member States which have started to realise their wind energy resource, renders the free trade of wind turbines across national borders difficult at best.

The publication indicates that market development subsidies backed by an effective RD&D programme and political goodwill, constitute the most effective way of developing wind energy markets.

Markets can be primed by:

- direct subsidy of the capital cost of wind plant;
- premium payments for the power produced;
- tax relief on wind plant investments;
- directives for access to the utility grid.

In order to succeed, however, these market primers need the backup of:

- harmonized standards for wind turbine construction and design;

- physical planning strategies including guidelines for the establishment of land zones for wind plant;

- active electric utility sector involvement and support;

- appropriate legislation, where necessary, requiring planning authorities and utilities to make adequate provision for wind power.

RD&D requirements have already been discussed. To summarize, the issues are:

- resource access: the need to get more output from each site - requiring RD&D on larger machines and terrain and wake effects;

- conversion efficiencies: the need to get more output from each machine - requiring RD&D on airfoils and controls;

- reliability and lifetime: the need for longer lifetimes and greater reliability - requiring RD&D on structural dynamics, fatigue and manufacturing;

- market compatibility: integration with utilities - requiring RD&D on utility integration issues (including isolated grids).

Wind power is technically mature and economically viable today ... at this stage it simply needs the chance to demonstrate its abilities. Like other new technologies, this requires establishment of the right conditions for its growth.

3 Solar Heating & Cooling and Daylighting

3.1 Technical Review of Solar Heating & Cooling and Daylighting Technologies

3.1.1 Current Status

3.1.1.1 Active Solar Heating and Cooling

Active solar energy systems may be employed in residential and commercial/industrial buildings for the provision of space conditioning (heating and/or cooling) and/or hot water. Another common application is the heating of swimming pool water.

The basic building block of an active solar energy system is the collector which contains a receiver or absorber that converts the incident solar radiation into collected energy. Energy collected is transferred to a working fluid for transport directly to the load using pumps, pipes and valves - or to insulated storage tanks for later use.

To meet daily loads during prolonged periods without sunshine, active solar systems can be supplemented by a backup conventional system.

The combination of solar heating with solar cooling capitalizes on year-round energy collection and utilization capacity and can therefore significantly increase the cost-effectiveness and energy contribution of solar installations.

Active Solar Heating:

Active solar heating is a well established technology - the main application being domestic hot water supply. Efficiencies depend on many factors including the collection temperature. In most solar heating systems, the solar collector either satisfies (at good insolation) the full load, or it can act as a pre-heater, taking a heat transfer fluid from a storage tank and heating it throughout the day. The heat transfer fluid is then heated by conventional energy to take it to the required demand temperature.

'Micro-flow' heating systems have excited considerable interest in the last few years and thermal performance improvements of up to 10% have been demonstrated (in microflow systems, the store is charged directly to enhance stratification (external heat exchanger)).

Active Solar Cooling:

Active solar space cooling is an area where the load and the seasonally available solar irradiation are well matched. Solar cooling system development, however, today lags considerably behind its heating counterpart. Notwithstanding this, work is underway on the development of a number of active solar cooling systems - but existing systems are expensive and commercialization is still a long way off.

At present, solar desiccant and solar-driven absorption systems are thought to have the greatest potential.

A number of absorption-type solar air-conditioning systems, tested in Europe were found to be technically feasible.Yet, although absorption systems are industrially well developed (from small systems such as refrigerators to very large cold-storage warehouses), the efficiency of solar heated evaporators with, say, flat-plate collectors, for example, is too low to ensure an economic return even with evacuated collectors. With such a low efficiency, the collector area becomes exceedingly large and costly.

In spite of this - and although the cooling market may remain limited in Europe, R&D work should continue as potential export markets are substantial and there could well be success with some absorption cooling machines - particularly those which work at low temperatures (LiBr/H_2O systems using a normal flat plate collector to supply the heat, for example). With regard to development prospects, solid-gas absorption-desorption systems should receive considerable attention.

Collectors and Transparent Insulation Technology:

Solar Collectors - the centrepiece of all low and medium temperature Solar Thermal applications, have undoubtedly become more reliable and more efficient in recent years.

Transparent Insulation Materials (TIM) and non-focussing concentrators - mainly as Compound Parabolic Concentrator (CPC), are currently used in addition, to improve conventional techniques of collector construction.

The major issue in collector development is production and installation cost reduction.

Flat plate collectors are the most commercialised at present. Flat plate collector technology is considered relatively mature and it is unlikely that there will be any major performance improvements. Cost reductions may result from improved manufacturing techniques and simplified system design.

High efficiency flat plate collectors are today using selective coating, a glass cover and a teflon sheet as a convection barrier.

A new bifacial flat plate collector with an absorber plate covered on both sides with TIM and combined with two spherical mirrors can reach an efficiency of 50% at ΔT=150 degrees at an insolation of 800 W/m^2.

An efficiency of 40% could be achieved under the same conditions with a normal flat plate collector if filled between absorber and cover with monolithic silica aerogel after evacuating the system to 0,1 bar. Slightly less efficient but most probably less costly is an evacuated flat plate collector with selectively coated absorber.

Evacuated tubular collectors are also technically mature. They are suited to mass production but at present are made on a batch basis only. There is little development work underway at present, apart from improved product development by the manufacturers.

Unglazed Solar Collectors are suitable for Heating at low temperatures only. They are made from polypropylene.

Airglass (silica aerogel) has been used in Denmark and translucent insulation material (PMMA-foam, for example) in Germany, with a view to reducing the heat loss from the transparent cover of collectors. [1]

There have been problems with both materials, however. Airglass is extremely sensitive to moisture, for example while the translucent insulation materials used were not yet stagnation proof and therefore could only be used in integrated collector-storage systems where temperatures never exceed 95° C.

More recently, however, a newly developed PMMA-foam transparent insulation material has been proven to be capable of withstanding temperatures of up to 160° C.

3.1.1.2 Passive Solar Heating and Cooling

A passive solar building works as an integrated system incorporating solar energy collection, distribution and storage, together with ventilation and auxillary heating.

Much progress has been made in the development of passive solar building designs and materials in the last few years, and many diverse elements, systems and designs have been combined with considerable success.

Passive Solar Heating:

The primary concern in passive solar heating, is the minimization of heat loss and the maximization of opportunities for useful solar gain without overheating.

Hence, the design of the microclimate around the building is an important concern in passive solar design.

---

[1] It is envisaged that silica aerogel will also play an important future role as a material integrated into double glazed windows.

Passive Solar Heating Configurations for the design of the building itself, include:

- Direct Gain Systems:
  consisting of large areas of south-facing glazing and constituting the simplest and easiest to build solar heating system;

- Indirect Gain Systems:
  which combine the collecting, storage and distribution functions within some part of the building envelope which encloses the living spaces (for example: the trombe, mass or water wall or the roof pond) - the performance of these systems can be considerably improved by cladding the outer surface with transparent insulation;

- Isolated Gain Systems:
  in which solar collection is thermally isolated from the living spaces of the building and energy transfer from the collector the the living space or storage and then the living space is effected by convection or radiation, etc.;

- Dual Gain Systems:
  designed to maximize on the main advantages of each system involved.

Passive Solar Cooling:

Passive cooling technologies have been in use in hot countries for centuries but modern building systems combined with the technical availability of active cooling, have meant that the traditional techniques have been largely forgotten. Passive cooling technologies remain, however, an important potential solution path for southern European countries confronted with the problem of maintaining summer comfort levels.

Research continues on the principal factors underlying passive cooling technologies, viz.:

- microclimate, siting and site layout
  (data, design and evaluation tools, etc.);
- building form and external finishes;
- building envelope
  (components, materials, 'leakage' level and coherence with climatic conditions);
- thermal mass;
- air movement;
- avoidance cooling (insulation, shading,
  thermal inertia, control of internal gains, etc.).

The principal Passive Solar Cooling Systems under ongoing development at present are:

- Ventilation Systems;
- Ground Cooling Systems;
- Evaporative Cooling Systems;
- Radiative Cooling Systems.

Several techniques and tools are now available and are beingused in major non-domestic demonstration projects.

3.1.1.3 Daylighting

Within the area of lighting, two distinct areas of development have emerged, viz.:

- control systems incorporating photo-electric controls;
- daylighting.

Control Systems incorporating Photo-electric controls:

Photo-electric controls switch off or dim unnecessary auxilliary lights when not required and are a well developed technology. Theoretical studies and practical tests have indicated that the saving potential with lighting control may be somewhere between 10% and 80 % depending, amongst other factors, on the type of building and the choice of control strategy. However, current knowledge of control strategies and systems and their implementation in different types of buildings is very limited - particularly with regard to combined control by solar control devices and auxilliary lighting.

Daylighting:

Daylighting is the use of natural sunlight to provide a building's lighting requirements.

Natural lighting may be provided directly to interior spaces or to spaces directly adjacent to the building exterior. The former case constitutes a Core System and the latter, a Perimeter System.

Technologies include advanced windows, light shelves, skylights, roof monitors and sidelighting.

An important area for development is the possibility of integrating lighting control systems with new technologies for directing daylight deeper into buildings and for solar control.

Atria designs are the dominant technology for core daylighting at present with current research focussing on new materials.

Thermally insulated daylighting elements can provide combined lighting and heating benefits - the transmitted irradiation illuminating the room before being absorbed in the walls and converted into heat.

The interaction of daylighting with heating and cooling loads must be considered and a necessary balance achieved - research continues on the development of appropriate design tools.

3.1.2 Development Prospects

The principal issues for Solar Heating & Cooling and Daylighting Technologies are:

- market compatibility;
- awareness.

The principal constraint on deployment is that technology is not reaching the marketplace - hence, the principal opportunities are continued R,D&D and technology transfer.

These issues are discussed in detail in Section 6.

3.2 Economic Review of Solar Heating & Cooling and Daylighting

It is almost impossible to make an assessment of the capital costs of solar heating & cooling and daylighting outside the context of a specific application.

With regard to passive systems, Steemers et al. (1992) observe that:

- additional costs for passive design vary from zero additional costs over normal design or cost yardsticks up to a 20% increase in extreme cases;

- zero or minimal costs are associated with planning and relocation of glazing, whereas sunspaces, trombe walls and so on, can be expensive;

- solar features are frequently included for reasons of amenity (particularly, sunspaces and larger areas of glazing) and thus the full costs cannot be attributed simply to the desire for energy saving.

Daylighting is a little or no capital cost item with large energy savings due both to reduced luminaire consumption and reduced cooling loads.

Passive cooling can be a zero or negative cost item if money saved by the ommission or reduction in size of active cooling systems (air conditioning) systems results.

In active heating and cooling systems, the principal cost determinants are:

- collector costs;
- rest of system (store, pipework, pump, controls, etc..) costs;
- installation costs ...

each constituting approximately one-third of the overall system cost. Maintenance costs can also be significant.

Direct energy savings are also difficult to assess.

Some application-specific costings follow. They are based on the recently published 'Solar Architecture in Europe: Design, Performance and Evaluation' (Steemers et al., 1991).

- Case One: The UK Giffard Park Housing Co-operative
  Development, characterised by:

    - careful site planning and orientation to allow
      maximum solar gains;

    - high levels of insulation and passive solar design
      features which reduced space heating requirements
      by 61%;

    - costs quoted for one terrace (1984/5) of:

whole terrace	-	UK£190,000
solar collectors	-	UK£1397
additional insulation and draught proofing	-	UK£8281

- Case Two: The Boegehusene Project - a short terrace of
  two storey houses in a suburb near Copenhagen,
  characterised by:

    - careful orientation and internal planning of the
      houses and the use of a sunspace;

    - a flexible plan of three 'climate zones' (south,
      middle and north) for the different seasons;

    - costs quoted:

building	-	ECU90,000
sunspace	-	ECU12-20,000

- Case Three: Les Basses Fouassieres - a development of
  twenty seven 3 and 4 bedroom family houses constructed
  in short terraces, characterised by:

    - a combination of passive solar features reducing
      heating bills by 37%;

    - careful design ensuring that all main rooms face
      south for direct solar gain - 90% of glazing is on
      the south facade;

    - a two storey sunspace with internal balcony
      providing a pleasant living space and warmed
      ventilation air to the whole house;

- trombe wall panels providing heated ventilation
  air directly to living rooms and bedrooms;

- costs quoted (1982 prices) of:

building (27 houses)	-	ECU1,558,526
average per house	-	ECU57,723
all solar features per house	-	ECU5,918
sunspace per house	-	ECU2,280
heat storage wall per house	-	ECU1,819
trombe walls per house	-	ECU1,588

- Case Four: The UK JEL Low Energy Production and
  Office Building, characterised by:

  - low energy and passive solar design to promote
    a high-tech business image and provide a focal point
    for the industrial park involved;

  - maximum use of natural light and low energy
    lighting with automatic controls to reduce
    electricity costs;

  - warm air provided to production area from glazed
    south facing wall and atrium;

  - use of blinds, louvres and natural ventilation to
    control overheating from solar gain ;

  - control over heating provided by zoning the
    building and the use of environmental controls
    and an energy management system;

  - overall quoted cost of UK(1983)£600,000

- Case Five: The Spanish Los Molinos
  Environmental Education Centre Project, characterised by:

  - having been designed to demonstrate a range of
    passive heating and cooling systems;

  - direct solar gain captured through large south facing
    windows with wooden shutters for shading in summer;

  - small ventilation openings on the north facade;

  - heating and cooling being provided by a new
    passive component: the 'blanco wall';

- use of sunspace corridor to allow direct gains to adjoining classroom and storage of solar energy in water drums;

- double glazed sunspace tower linking ground and first floors, housing a staircase surrounding a water tank to store solar gains;

- a central patio area which remains open to cooling easterly breezes in summer;

- a fountain in the patio which humidifies the surrounding air - thereby assisting cooling;

- use of high levels of insulation in walls and roof;

- supply of all hot water requirements by means of an active solar water heating system;

- comfortable temperatures achieved throughout the year without auxiliary heating or cooling;

- innovative design successfully integrated with local rural architecture;

- costs quoted as:
  building   -   ECU189,394
  solar features   -   ECU8,661

- Case Six: The Vielha Hospital - Spain, characterised by:

  - clerestory windows at the top of the south facade to supply the north side of the top storey with natural daylight;

  - careful facade design with a large area of south facing glazing and shading to optimize useful solar gains control summer overheating;

  - massive walls and floors to provide efficient heat storage;

  - a compact shape, resulting in a low surface to volume ratio, high levels of insulation and a draught lobby to minimize heat loss;

  - overall heating requirements reduced by 62% compared to a conventional hospital design;

  - passive solar gains contributing 33% to the gross

space heating load;

- natural ventilation in summer, achieved by opening windows in the north and south facades, providing the necessary cooling in most areas, reducing the need for air conditioning except in specific areas;

- zoning control systems to maximize the use of solar gains and minimize consumption of auxiliary fuel;

- costs quoted (1985 prices) of:

building	-	ECU2,920,000
solar features	-	ECU22,300
extra wall insulation	-	ECU4,700
controls for heating system	-	ECU7,300

3.3 Environmental Impacts and Public Acceptability of Solar Heating & Cooling and Daylighting, in the European Context

The principal environmental impact / public acceptability issues relating to solar heating & cooling and daylighting in Europe are:

- energy costs of materials;
- pollution reduction;
- avoidance of 'sick building syndrome';
- architectural integration.

(i) Energy Costs of Materials

Energy payback times vary with technology, application and materials.

In the case of some active heating and cooling systems - and, in particular, for passive cooling systems, the energy payback time may vary from one month to many years depending on the climate and on the intensity and duration of use.

For example: the energy payback time of a solar panel is about a year - while an extra area of double glazed south facing window will collect its energy cost in one month.

(ii) Pollution Reduction

Taking passive solar design, for example, DGXII's 1990 study 'Passive Energy as a Fuel 1990 - 2010' remarks that the reduction in the use of fossil fuels brought about by the current usage of passive solar energy prevents:

- 229 million tonnes of CO_2
- 1.3 million tonnes of SO_2
- 0.56 million tonnes of oxides of nitrogen NO_x ...

being emitted to the atmosphere each year.

If the technical potential solar contributions were achieved the amount of CO_2 saved by the year 2000 would rise to 272 million tonnes. Legislation aimed at reducing SO_2 and NO_x emissions from power stations generally would, of course, mean that there would be less scope for reduction of emissions of these gases from the use of passive solar design.

(iii)   Avoidance of 'Sick Building Syndrome'

Up to 50% of Primary Energy Use in Europe is in buildings.

As DGXII's 1990 study 'Passive Energy as a Fuel 1990 - 2010' indicates, passive solar design can help to alleviate 'sick building syndrome' which consists of, for example:

- low fresh (outside) air ventilation rates;
- low air movements;
- some types of heating and ventilation systems;
- fluorescent lights (glare and flicker);
- proportion of daylight versus artificial lighting;
- non-openable windows;
- tinted or mirrored glass ...

and which can result in low user moral, absenteeism and reduced productivity.

Passive design uses more daylighting, elimination or reduction of air-conditioning, natural ventilation and greater use of passive solar heating - thus putting building users more closely in touch with the outside environment.

(iv)   Architectural Integration

It is evident that good architectural integration is essential if there is to be large-scale deployment of active and passive solar heating, cooling and lighting systems.

As Tributsch (1992) observes, efforts must be made to establish an aesthetic form for solar structures in which their active and passive systems fulfil their structural and energy-saving purposes, while at the same time creating a cohesive, expressive artistic expression.

Obstacles to the wider deployment of solar heating & cooling and daylighting systems in Europe include: ignorance regarding available technologies, lack of demand on the part of owners and designers, fear of problems and failures and non-availability of design tools, materials and products.

Such obstacles can be overcome, however, through, for example: education, publicity and promotion, building regulations, grants and other incentives.

3.4 The Commercialization of Solar Heating & Cooling and Daylighting in Europe

Figure 4 gives an overview of the contribution of solar energy to heating, cooling and lighting buildings in the EC in 1990 and the year 2000. Figures are shown for a base case of no action taken to increase solar use and for the case of thorough use of solar potential.

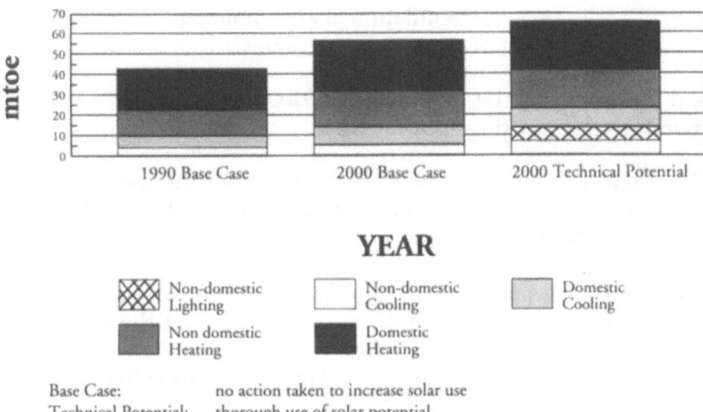

Fig. 4. Contribution of solar energy to heating, cooling and lighting buildings in the EC in 1992 and 2000

While photo-electric controls are becoming a reasonably well established technology within the Community, considerable development and demonstration is still required to establish daylighting widely in new building design and refurbishment.

With regard to active solar heating/cooling technologies, 50% of solar panels installed in the Community at present, are located in Greece, while France and Italy also have relatively high levels of implementation.

Those parts of the community which are characterized by climates less favourable to active solar heating/cooling technologies, could, however, in addition, meet up to 50% of the heating requirements of new buildings, from solar energy if efficient passive-solar components were used.

Solar heating and cooling subsectors currently economically viable include:

- active heating (depending on location);
- passive heating;
- passive cooling;
- active and passive domestic hot water heating...

specifically, in regard to the following applications:

- actively and passively heated swimming pools;
- heating for new and existing buildings;
- cooling for new buldings;
- domestic hot water;
- agricultural applications, viz.:
  - solar dryers;
  - natural ventilation in greenhouses;
  - frost protection in greenhouses ...

(all low and medium temperature - further RD&D is needed for high-temperature solar heating applications).

About 25% of Europe's total energy consumption is for the space conditioning of buildings - thus there is a very large potential market for solar heating and cooling technologies in Europe.

It is estimated that passive solar technologies alone, if combined with energy conservation strategies, could save more than half of all the energy currently consumed in this sector - yet passive buildings are still not widely constructed within the Community ... the technology is simply not getting to the marketplace.

Developers, owners of buildings and users of buildings want artificial features - they do not realise that passive buildings perform better, use less energy and are liked more by and healthier for their users. The problem is that no formal mechanisms exist to allow the designer, developer, owner, estate agent, user to assess and promote the benefits of passive design.

An assessment and promotion scheme which gives due credit for passive low energy design, features and performance should be developed and used throughout the Community (possibly with the assistance of National Governments' endorsements of the scheme through the SAVE program) in conjunction with a central certification body and approved assessment associate organizations.

The United Kingdom's BREEAM scheme (developed jointly by ECD and the Building Research Establishment) could be used as a basis for the scheme, the development of which, to fit individual Member State requirements, would necessitate consideration of context-specific:

- heating, cooling and lighting energy calculations;
- 'sick building' indicators;
- installation of active solar and photovoltaic technologies;
- other environmental considerations as appropriate;
- examples of good practice.

Benefits of such a scheme would include the following:

- good buildings would be given a high profile;
- building construction, fuel running and maintenance costs would be reduced;
- increased productivity of building users;
- contribution to environmental improvement.

Passive solar buildings represent a particularly attractive solar heating/cooling/daylighting - and, indeed, renewable energy - option because many materials and techniques can be incorporated into new buildings at little additional cost. Increase in awareness is, however, essential to the increased deployment of solar buildings technologies.

Currently, the estimated share of Community energy consumption by the year 2000 for solar heating & cooling and daylighting technologies - active and passive, is 1-2% (an increase in R,D&D and market stimulation activities as outlined in Section 6 of this Report could, however, significantly increase this figure).

4 Photovoltaics

4.1 Technical Review of Photovoltaics (PV)

4.1.1 Current Status

Photovoltaics is the direct conversion of sunlight into electricity using devices made of thin semiconductor layers.

Figure 5 gives an overview of Europe's solar energy potential.

Fig. 5. Europe's annual daily global irradiation in KWh/m²: mean annual means 1966–975

PV Modules:

The basic element of a PV system is the solar cell which produces electrical energy. A PV module (typically 0.3 to 1.00 m^2 in size in the case of commercially available modules) consists of a number of solar cells connected together.

The peak output power (Wp) of a module (defined as the power delivered at an irradiance of $1000W/m^2$ @ 25°C and AM 1.5) generally ranges from 30 to 120W. [1]

Groups of modules can be connected together to form larger fields and by means of this modular extension PV systems of any size can be achieved.

There are two principal types of PV module:

-       the flat plate module
        in which the whole irradiated area is covered with
        solar cells;

-       the concentrator module
        with its optical elements (mirrors, lenses) which
        concentrate the incident light onto a small area
        equipped with solar cells.

Flat plate modules convert both direct and diffuse solar radiation while concentrators must have direct radiation to work effectively. Thus while the former may track the sun (giving increased energy yield), the latter must.

Solar Cell Types:

PV cells can be categorised as crystalline or thin film. The cells can operate with a lens or mirror to form a concentrator.

Currently commercially available photovoltaic cells are:

-       monocrystalline silicon:
        with laboratory efficiencies of 23% for normal cells and 28% for
        concentrator cells - the corresponding figure for commercial
        modules (normal cells) is 15%;

-       polycrystalline silicon:
        with laboratory efficiencies of 18% for normal cells   - the
        corresponding figure for commercial modules is 12%.

---

[1] There is already a trend towards larger custom-made modules of up to $2m^2$, >200Wp.

- thin film:

with laboratory efficiencies of:

13%	for	amorphous silicon (normal cells);
25.5%	for	gallium arsenide (normal cells);
15%	for	CuInSe2 (normal cells);
15%	for	cadmium telluride (normal cells);
29%	for	gallium arsenide (concentrator cells)
>10%	for	TiO_2 organic cells ...

at present, the only commercial thin film cells are amorphous silicon, for which commercial module efficiencies are 5%.

In summary, mono- and poly- crystalline cells are generally characterized by high efficiencies and great robustness and are considered technically mature though there is still scope for improvement in areas of materials and manufacturing techniques.

The main applications for mono- and poly- crystalline cells will be in roofs and central power systems and the less efficient (but lower cost/Wp) amorphous Si cells will be used in building cladding (for which a large potential market exists (Hill, 1992)) and in consumer products such as calculators, watches, etc..

Finally, multi-junction cells, by utilizing a larger portion of the solar energy spectrum, can yield higher efficiencies than single-junction cells - these are, however, still at the laboratory development stage.

Concentrator Cells:

The main application for concentrator cells would be centralised grid-connected systems in areas with a very high (>90%) level of direct sunlight.

Higher efficiencies (over 30%) can be obtained as well as a reduction in module cost.

Some demonstration plants in the USA have shown good performance - however, the need for moving parts has prevented their more widespread use and the increase in efficiency does not offset additional lens and mirror costs.

Fig. 6. The PV installation at Fota Island, Cork, Ireland

Solar Generator Characteristics:

Flat modules are currently available on the market. Concentrator modules are not.

Tests on presently manufactured modules based on crystalline silicon show lifetimes of at least twenty years with no detectable electrical degradation.

Limits of lifetime are attributed to corrosion phenomena of the module materials glass, metal and plastics.

Module replacement rates (necessitated by failures, fractures or other mechanical damage) is about 0.2% per annum.

In the case of amorphous silicon modules, there is the additional problem of light induced degradation which causes a reduction in the cell efficiency with exposure time. After the first few hundred days of operation the efficiency reduces to approx. 5% which is 65-75% of the initial value. This low efficiency clearly restricts the application of amorphous silicon modules in large PV power stations.

Interestingly, thin film materials such as CIS demonstrate much higher stability. In consequence great efforts are currently focussed on bringing these latter materials to early market maturity.

Other important characteristic design parameters for the construction of PV systems include: module size, weight, mechanical stability and the manner by which it is secured to its support.

Other PV System Components:

Other PV system components are conventional module support structures and the electrical components for power conversion and conditioning - the design of both being based on the size and type of PV system applied.

Inverters:

In applications where PV systems generate AC power the second essential component after the solar generator, is the inverter.

Despite the fact that the inverter represents a conventional component, there is still much development effort required here. Conventional inverters which use thyristors switched at low frequencies, exhibit decreased conversion efficiencies (<90%) and higher losses when operated in the partial load region - and, as PV systems operate most of the time in the partial load region, the efficiency of conventional inverters averaged over longer time spans is low: typically, 70-80%.

Another significant problem with conventional inverters is the level of harmonic distortion produced (>30%) and power factors <0.8.

The development of new types of inverters (employing high speed MOSFET and insulated gate bipolar transistor switching circuits) specifically for PV systems addresses this problem. These have efficiencies of between 90 and 95% over an operating range of 10-100% rated output.

Other areas of power conditioning:

In the other areas of power conditioning less development is required as conventional elements and components perform adequately.

In the case of large PV installations which cover large areas of land, significant cost factors are: DC wiring, DC/AC switching and protection and, especially, lightning protection.

Support Structures:

Much work is still required in the area of PV support structures as related land requirements are considerable and the costs of solar generator support structures are high. As these structures must often withstand high wind and snow loads, they must be designed and constructed accordingly.

Work carried out to date on support structure optimization has indicated cost reduction potentials of up to 40% (Helm, 1992, personal communication).

As costs of support structures are strictly proportional to module surface area, higher module efficiencies in the future will proportionally reduce land requirements.

Work is also currently underway to investigate the possibility of replacing massive support structures with lightweight or stretched cable structures and on the integration of PV modules into existing structures (roofs, etc.).

Grid-Connection:

Grid connected PV plants must, of necessity, meet specific requirements with regard to the quality of the supplied electricity (voltage, frequency, harmonics content, total harmonic distortion) and with respect to safety measures and generally accepted rules and standards.

This leads to higher material specifications as certain design standards for the electrical system components are necessitated.

More generally, the performance of grid connected systems is well documented.

Various investigations show that in southern European countries daily average outputs of 4kWh/kWp can be achieved - with annual availabilities in excess of 94% and load factors of about 20%. In north-western Europe daily outputs of 2kWh/kWp and load factors of about 11% can be expected.

Stand-Alone Applications:

PV cells also have a wide range of stand-alone applications, i.e. not connected to the electricity grid.

These applications include, for example:

- power supplies for remote dwellings and villages;
- vaccine refrigerators;
- sea-water desalination;
- water-pumping;
- consumer products such as calculators.

The power supplied for these applications varies from a fraction of a watt for a calculator to, for example: 1-2 kilowatts for a remote water pump and up to tens of kWs for water desalination.

The reliability of stand alone systems must be high. In fact, for most of these applications, the reliability of the system is at a premium over its cost - and this is one of the reasons why PV technology is most suitable when combined with compatibly reliable conventional systems technology.

Fig. 7. A PV-powered factory at Oberburg near Burgdorf, Switzerland

System Performance:

Experience to date with PV systems has shown that, in general, systems performance problems may be due to:

-   hardware failures (cracked modules, inverter problems and short battery life) and, in particular, inadequate hardware backup and spare parts;

-   non-optimized hardware design;

-   non-optimized power management - especially with regard to batteries;

-   incorrect sizing of components - particularly the inverter;

-   changes in load demand;

-   changes in load profile.

l.1.2 Development Prospects

Table 2 summarizes the principal current technological constraints and opportunities for Photovoltaics, based on the 1990 U.S. Interlaboratory White Paper SERI/TP-260-3674 DE90000322 on the Potential of Renewable Energy.

Table 2. The Principal Current Technological Constraints and Opportunities for Photovoltaics.

Issue	Constraint	Opportunity
Conversion Efficiency	Efficiencies are not high enough	R,D&D in materials science, inverters and power conditioning equipment
Reliability/ Lifetime	Proof of lifetime needed	Long-term demonstration
Market Compatibility	Mass production of package systems	Some standard designs and sizes
Market Acceptability	Valid and reliable rating criteria	New approaches to PV plant rating
Manufacturability	Need to develop low-cost process and to prove scaled-up manufacturing	Publicly funded process R,D&D and demonstrations

The issues identified in Table 2 are discussed in detail in Section 6.

4.2 Economic Review of Photovoltaics

The Review of Present and Predicted Future Performance and Costs of Photovoltaics in Europe presented here is based on analyses by Zweibel (1990), Palz and Schmid (1990) and Mertens et al. (1992).

4.2.1 Cost Determinants

PV system costs can be broken down into two major categories, viz.

- PV module costs;
- Balance of Systems (BOS) costs.

(i)   PV module costs

Module costs vary with module type and, of course, production volumes.

For quite some time now forecasts of low cost (<1.0 ECU/W_p) PV cells have been made - but have not materialised to date.

Current and year-2000 production cost estimates made at the 1992 (11th.) European Photovoltaics Conference Workshop as a result of a survey of European suppliers (see Mertens et al., 1992), were as follows:

- for Crystalline-Si Modules:
  3 ECU/Wp now and 1.8 ECU/Wp in the year 2000
  (assuming a cell efficiency of 13%, a module lifetime of 20 years and a cell size of 100 cm^2 now and a cell efficiency of 16%, a module lifetime of 30 years and a cell size of 225 cm^2 in the year 2000);

- for Thin Film - a-Si Modules:
  2.1 ECU/Wp now and 1.0 ECU/Wp in the year 2000
  (assuming a cell efficiency of 5% now and a cell efficiency of 10% and a module size of 5000 cm^2 in the year 2000);

- for Thin Film - CdTe Modules:
  1.5 ECU/Wp in the year 2000
  (assuming a cell efficiency of 15%, a module lifetime of 20 years and a module size of 5000 cm^2);

- for Thin Film - CIS Modules:
  1.5 ECU/Wp in the year 2000
  (assuming a cell efficiency of 15%, a module lifetime of 20 years and a module size of 5000 cm^2).

The schematic shown in Figure 8 - Mertens et al.'s (1992) 'PV-Module Learning Curve', demonstrates the effects of increased production volumes on module costs.

Deflated market prices ($(1990)) are plotted against cumulative volumes and show that (assuming an average inflation rate of 6%) as production volumes double, market prices are reduced by 17.5%. (Note that $(1990)3.6 is equivalent to 2.8 ECU(1992).)

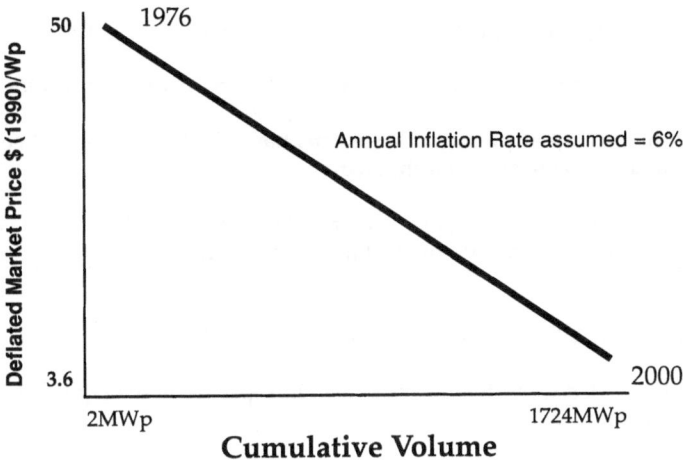

Fig. 8. PV module learning curve

(ii) *Balance of Systems (BOS) costs*

Balance of Systems costs incorporate the cost of everything other than the modules. BOS costs can therefore include:

- system design costs;
- land and site preparation costs;
- system installation costs;
- the cost of support structures and their foundations;
- power-conditioning equipment and related electrical system costs;
- operation and maintenance costs;
- indirect costs (contingencies);
- fixed costs (interest expenses for capital);
- storage and related costs where applicable.

The principal BOS costs are those relating to the inverter, power conditioning, frames, support and wiring.

Current and year-2000 cost estimates (ECU/Wp) made at the 1992 (11th.) European Photovoltaics Conference Workshop based on a survey of European suppliers, were as follows (see Mertens et al., 1992):

		1992	2000
-	inverters	1.7	0.6
-	power conditioning	0.8	0.4
-	frames, support and wiring	1.0	0.8
-	TOTAL BOS (no batteries)	3.5	1.8

Note: These are mean figures and are not specific to a particular application.

It should be noted that at present, PV systems have the highest capital costs per MWh/yr of all renewable energy technologies.

PV systems have proven reliable, however, and O&M costs are low - most costings assume maintenance of about 1% of capital cost per year.

4.2.2 Overall Installation/System Costs

Total current Installed Costs for PV systems in Europe, are in the region of 6 - 8 ECU per peak Watt.

Overall system prices per installed power unit (based on large production volumes) are approximately:

- 6.5 ECU per peak Watt for the simplest case of a module integrated into the roof of a grid-connected building without storage [1]

  (assuming a module cost of 3 ECU per peak Watt and inverter/power conditioning/wiring and support costs of 3.5 ECU per peak Watt) ..

- 7.4 ECU per peak Watt for large PV Power Stations

  (assuming a module cost of 3 ECU per peak Watt, land/support structure costs of 1.9 ECU per peak Watt and inverter/grid-connection costs of 2.5 ECU per peak Watt).

  Power stations are generally more expensive than systems installed on buildings as additional costs for land, site preparation and support structures, etc., must be incurred.

---

[1] Results of a survey of European suppliers - it should be noted, however, that prices of 11.5 ECU/Wp characterized the recent '1000-Roof Programme' in Germany, while equivalent Swiss programme prices were slightly less.

Year-2000 cost estimates (ECU/Wp) made at the 1992 (11th.) European
Photovoltaics Conference Workshop, for grid-connected systems, were as follows:

		2000
-	modules	1.8
-	BOS	1.8
-	TOTAL	3.6

kWh Cost Estimates

Finally, current and year-2000 ECU/kWh cost estimates made at the 1992 (11th.)
European Photovoltaics Conference Workshop, for grid-connected systems, were
as follows (see Mertens et al., 1992):

	1992	2000
W-Europe (1100 kWh/m^2):		
- no avoided costs	0.7	0.4
- avoided costs: (70 ECU/m^2 or 0.5 ECU/W)	0.6	0.3
S-Europe (1750 kWh/m^2):		
- no avoided costs	0.5	0.3
- avoided costs: (70 ECU/m^2 or 0.5 ECU/W)	0.4	0.2

4.3 Environmental Impacts and Public Acceptability of Photovoltaics in the European Context

The principal environmental impact / public acceptability issues relating to photovoltaics in Europe are:

- energy costs of materials;
- safe handling, storage and disposal of materials;
- land requirements and impacts on wildlife and natural habitat;
- safety.

(i) Energy Costs of Materials

Most analyses of the energy costs of materials for photovoltaic systems focus on the energy content of PV modules.

In the case of Polycrystalline Silicon Modules, about half of the energy invested in the module is associated with wafer manufacturing. Energy payback times for crystalline silicon modules have been estimated by Palz and Zibetta (1991) to be between 1.6 and 2.7 years, depending on radiation levels.

In the case of Amorphous Silicon Modules, the glass substrate is the object of highest energy expenditure, followed by the SnO_2 layer and a-Si deposition. The average energy payback time for this cell is 1.3 years under typical European climatic conditions (again, Palz and Zibetta (1991)).

With concentration systems, the payback time is somewhat shorter (less than a year) because of the reduced cell area.

Balance of systems (that is: those additional devices required to complete an operational power system) energy costs include those costs associated with inverters and support structures - the latter varying considerably depending on materials used.

(ii) Safe Handling, Storage and Disposal of Materials

During their operation, photovoltaic systems discharge neither gaseous nor liquid emissions, nor heat.

The production of photovoltaic solar generators can, however, involve hazardous gaseous, liquid and solid substances (feedstock gases used to make a-Si devices, for example, include silane which is highly toxic and explosive).

Negative environmental impacts could result from the incorrect handling, storage or disposal of these materials.

Experience with these materials in related industries has shown that application of the available controls can reduce potential hazards from these materials satisfactorily.

In order to minimize environmental impacts, great care must be taken in the disposal of toxic materials from photovoltaic manufacturing - and in the disposal of solar cell modules at the end of their useful lifetime. This is especially important in the case of CdTe and CIS cells.

(iii)   Land Requirements and Impact on Wildlife and Natural Habitat

Photovoltaic land requirements vary with application but become a significant issue mainly in relation to large-scale power supply.

In the long-term, it is expected that photovoltaic power will, to an extent, replace conventional power stations.

There are two basic means of large-scale photovoltaic power supply:-

> - firstly, and perhaps, conceptually, most feasibly, large
> centralised grid-connected plants (concentration schemes
> could be considered for this application);
> - secondly, the use of many small, roof systems.

The land requirements of large, centralized grid-connected PV plants are substantial and are much larger than existing centralised generation systems. Concomitant impacts on existing land use, ecosystems and habitats are to be expected.

Generally speaking, land requirements are up to three times larger for two-axis trackers than for single-axis or fixed arrays. This is because of the way that trackers move and potentially shadow each other when the sun is low in the sky.

The use of many, small roof systems cuts the land requirement to zero - though it is clear that exceptional architectural integration is essential if thousands of buildings are to be equipped in this manner.

(iv)   Safety

The safety record of PV energy technology is generally good.

A number of accidents have, however, occurred during manufacture, installation and use (electric shock and public fires, for example) - though these tend to result from poor management or non-observance of safety regulations, rather than from technical faults.

Photovoltaics is one of the most environmentally benign energy technologies and it has received considerable approval and support from the informed public because of its clean, silent and trouble-free operation.

4.4 The Commercialization of Photovoltaics in Europe

Current installation figures for the Community stand at approximately 150 MW - most of it in small decentralized units in the 1kW range, with about 20% of the capacity in larger units in the range 100kW-3MW.

The commitment of European industry to photovoltaics has strongly increased of late - yet overall markets are still very modest in size. The current European PV module shipment is about 12MW/year and the world shipment rate is over 50MW/year. By 1995 the European PV manufacturing capacity will be 24MW and the world capacity will be 154MW. Figure 9 gives a percentage breakdown of 1990 European/World module shipments by cell technology.

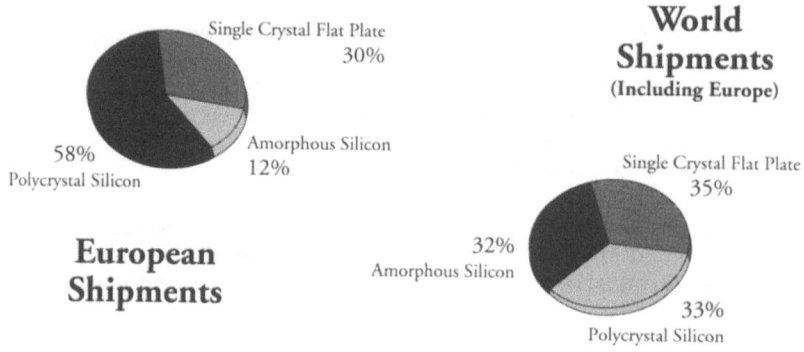

Fig. 9. Percentage breakdown of 1990 European and world module shipments by cell technology

Notwithstanding the fact that existing uses of PV are modest, they are significant. The market for individual social groupings is particularly important from a socio-economic perspective - over 4100 remote houses in Italy corresponding to 1.85MW have been electrified by installing photovoltaic power systems. There has also been PV electrication of 5 remote villages in Spain totalling 23.9kW.

The development of central photovoltaic power plants in the 300-3000 kW range for Europe's electricity grids is also being actively pursued in Italy, Spain and Germany.

Photovoltaic applications which are economically viable at present cost levels (6-8 ECU/W peak) are those for electricity supply to isolated sites - not connected to a grid. They include:

-   individual houses;
-   groups of houses / small villages;
-   sea-water desalination, water pumping
    and irrigation;
-   vaccine refrigerators;
-   lighthouses.

Factors favouring the further development of photovoltaics include the potential benefits of:

-   improvement of living conditions
    and infrastructures
    in the less favoured regions;
-   positive effects to counter rural depopulation;
-   environmental benefits ...

and, of course, prospects for further reducing installation costs - thus increasing competitiveness.

A striking feature of PV which places it in a very favourable market position, is its modularity. Indeed, PV is probably the most modular source of electric power, allowing:

-   systems to be employed in a wide range
    of sizes (<100W to several MW);
-   relevant, scalable experience to be gained
    with small, affordable systems;
-   short construction times
    (e.g. <1 year for 100kWp plant);
-   revenue generation as modular segments
    come on line ...

and thus reducing those technical and financial risks generally associated with testing pre-commercial deployment and commercial use.

It should be noted, however, that unless the current annual rate of photovoltaic installations increases substantially, the estimated photovoltaics share of the Community's energy consumption by the year 2000 will be marginal.

PV needs to be applied now to those ever-expanding niche markets where it makes economic sense today.

Exploitation of such niche markets can build PV experience for utilities and also create the volume of demand necessary to enable significant movement along the technological learning curve - and reduce costs.

In turn, reducing costs can further expand the size of the distributed utility niche market, incrementally moving photovoltaics closer to a future competitive position in central station bulk power markets ... and again, in turn, stimulating further technological developments - perhaps, for example, in the area of PV concentration technology.

5   Biomass

5.1   Technical Review of Biomass Energy

5.1.1 Current Status

5.1.1.1 Biomass Sources and Production and Pre-Processing Technologies

Figure 10 gives an overview of 1992 and year 2000 EC biomass resources.

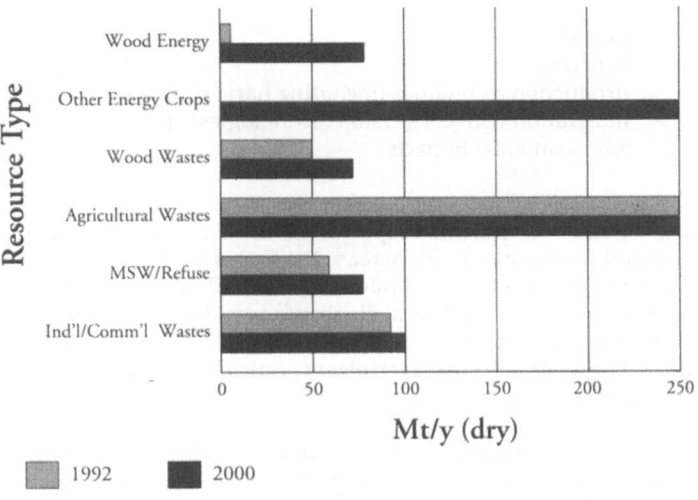

Fig. 10. European Biomass Resources (Mt/y (dry))

With regard to existing biomass sources:

-   wood from existing forests and coppices is, today, the main biomass source in Europe;
-   straw from cereal constitutes a lignocellulosic source of biomass;
-   sugar beet and cereal (wheat, barley, etc.) are used for ethanol production;
-   rapeseed is used for oil production.

Emerging sources - or 'energy crops' include:

-   perennial crops - a lignocellulosic source, with short rotation forestry - species including:
-   poplar in general;
-   willow in northern Europe;

- eucalyptus in west and northern Europe;
- robinia in marginal land.
- agricultural annual crops, viz.:
- sorghum - lignocellulosics/sugar,
  for mid and southern Europe;
- miscanthus (eastern Europe),
  cynara (dry, southern Europe).

Of course, for each of these species, further research is needed on:

- physiology;
- genetics;
- production techniques (including harvesting and
  integration with conversion technologies) and
- environmental impacts.

It is envisaged that developments in species selection and cloning and in planting, maintenance and harvesting could increase the Community's Biomass crop yields from their present development and pilot stages of 9-20 dry tonnes per hectare per annum (depending on crop and local circumstances) to yields of up to 15-35 dry tonnes per hectare per annum - and concurrent work on key species, to both increase productivity and decrease fertilizer/pesticide requirements, would have positive environmental effects.

Biomass is typically a diffuse, moist, low density fuel with an energy value of about 20% of that of oil or coal on a weight basis and available in many regions only on a seasonal basis.

This creates problems with regard to collection, transport and storage which are then compounded by the fact that wet biomass is subject to microbial decay and attack by insects. This can lead to loss of heat value and health and fire hazards.

Considerable work has been carried out on the collection, transport, storage, densification and upgrading of raw biomass to provide drier, more stable, energy dense fuels suitable for transport and storage, compatible with various engines which can be used as prime movers for electricity generation.

Recent developments include harvesters which combine the cutting, disintegration, storage and transportation of forest biomass and equivalent machines are now being developed for sorghum and short-rotation poplar.

5.1.1.2 Biomass Processing / Conversion Technologies

Virtually all Biomass needs to be processed in order to produced the primary energy sources that are ultimately converted into final energy. The specific processing/conversion technologies used depend on the state of the organic matter to be processed.

(i) **Biochemical Conversion:**

The principal Biochemical Conversion Technologies are:

- Acid- and Enzyme- based Hydrolysis;
- Fermentation / Distillation;
- Digestion;
- Pulping;
- the newly discovered lignocellulosic/synthesis gas/ethanol mixed thermal-biological route.

Acid- and Enzyme- Based Hydrolysis:

Hydrolysis is essential for the biological conversion of ligno-cellulosics.

During the 1980s, considerable efforts were directed towards research and development of acid-based systems and limitations identified as mainly related to engineering requirements and economics. In contrast, limitations to enzyme-based systems reflect the activity of current enzymes and costs of production.

Thus Acid-hydrolysis is already well established at laboratory scale while Enzyme-hydrolysis still requires substantial fundamental research.

Pre-treatments such as size reduction and the use of organic solvents have been investigated with a view to increasing the ease of hydrolysis. The process which appears of most value on a cost-benefit basis is steam explosion in which the woody material is treated with steam under pressure and then exploded by rapid reduction to atmospheric pressure.

Fermentation/Distillation:

Sugar- and starch- based systems are commercially available at present. Mass yields are in the region of 50% of sugar feedstock. Energy yields are more difficult to calculate but estimates are positive (see Wright Commission Report).

Where energy for the process comes from combustion of the fibrous part of the crop (as is the case with sugar cane) there is a net energy gain.

A wide range of novel yeasts, bacteria and fungi capable of producing ethanol from a wider range of raw materials have been identified and investigated in recent years. Of these, the bacteria Zymomonas mobilis and Clostridium thermocellum have received particular attention and have been used in a number of pilot studies.

A second fermentation (used commercially in the past) is that catalysed by another species of Clostridium (C.acetobutylicum), which yields a mixture of acetone, butanol and ethanol. Research here has aimed at improving the process by increasing the yield and increasing the resistance of the bacteria to inhibition by solvent. So far no major breakthrough has been achieved, however.

Digestion:

Biogas can be produced from any organic material when enclosed in an anaerobic environment. This process has been widely developed to use a wide range of wet biomass wastes and residues and as a means of waste treatment. The composition and the quantity of gas produced varies with the feedstock composition and the digester design. Generally; between 0.2 and 0.3 m-cubed of biogas are produced per kg of dry solids.

It is interesting to note that today, approximately 80% of world biogas 'production' originates from commercially exploited landfill as landfill gas.

To date, considerable research effort has been concentrated on, for example, factors affecting microbial population growth on start-up, the stability of systems on shock loading, temperature, mineral nutrients and inhibitors.

Currently the main focus of interest is on high solids digesters optimized for high volumetric gas production - and on high rate digesters aimed at rapid treatment of large volumes of dilute effluents from agro-industrial processes.

Pulping:

One of the newer pulping processes permitting the utilisation of conventional biomass including annual crops instead of slow growing traditional wood is known as 'The ASAM Process'.

It is a non-polluting technology which involves a closed water cycle and avoids all chlorine emissions and this activity would be integrated with the production of compost and organic fertilisers by anaerobic digestion.

(ii) Thermochemical Conversion

Thermochemical processes entail either temperature-related oxidation (slow-, conventional- or flash- pyrolysis; liquefaction; gasification; combustion) or purely chemical reactions (esterification).

The most promising thermochemical conversion technology currently seems to be the production of pyrolytic oil or bio-oil. Related technologies are evolving rapidly with improving process performance, larger yield and better quality products.

The economic viability of this process would seem promising in the medium term and its integration into conventional energy systems presents no major problems.

Pyrolysis:

At present the major pyrolysis product is charcoal. This can be used in small gasifiers, generating low energy gas suitable for use in spark ignition-engine driven generators in the kW range.

Other Pyrolysis products are bio-oil and fuel gas.

Current research efforts are concentrating on improving production of pyrolysis oil as a means of generating a higher energy density product at small local stations with the oil transported to central power stations in the MW range.

Liquefaction:

The high cost of pressure reactors, the need for feed preparation and unresolved problems of feeding slurries at high pressure as well as separation of product from solvent has reduced interest in Liquefaction to some extent of late - though basic research on batch reactors and catalytic hydro-cracking continue.

Gasification:

Low-heat-value gaseous fuels can currently be produced by the gasification of biomass in air. A wide range of small scale air-blown gasifiers of updraft, downdraft or more complex design have already been built and marketed - mainly producing fuel for engines (transport or electricity generation).

Work is underway, however, on the development of large (1000 tons per day raw material input) oxygen blown systems to produce higher value gas which may be catalytically upgraded to hydrocarbons or methanol or, alternatively, used as feedstock for fermentation - and on hydrogasifiers.

Combustion:

Grate and boiler design are the principal determinants of the completeness of combustion and efficiency of heat transfer. Direct combustion of biomass for electricity production is only effective as the primary process in fairly large systems for which a wide variety of grates, furnaces, suspension burners, pile burners and fluidised-bed combustors have been designed and used in the forest and food products industries to produce power for their own use. Increases in efficiency has of late resulted in an ability to export electricity to the local grid, however.

The largest biomass fuelled boilers use grate systems adapted from coal-fired systems. These may be up to 100 MW thermal, producing about 200 tons of steam per hour. Design considerations include methods of stoking, supplying combustion air and mechanical conveyance of the fuel from store to boiler.

The main development is in the increasing use of fluidised bed combustors as an alternative to grates. These have a high thermal efficiency and can burn a mixture of fuels of varying size and with up to 60% moisture content.

They are, however, limited in maximum steam pressure and steam temperature as well as size, below 35 MW thermal.

Esterification:

Recent developments include the successful testing of fuels containing between 5 and 100% of methyl or ethyl esters derived from vegetable oils in diesel-powered farm machinery and private cars. Pilot esterification plants designed to produce fuels from rapeseed oil have already been developed in France and Austria.

5.1.1.3 Biomass Systems

(i) Commercial Systems

Commercial Biomass Systems today include:

- Traditional Fuels;
- Forestry and forestry residue Fuels;
- Dry agricultural residues;
- Municipal Solid Wastes;
- Landfill Gas;
- Other wet organic wastes (WOW);
- Combustible industrial wastes;
- Ethanol.

Traditional Fuels

The technologies used range from open fires with very low efficiency (10-25% of heat utilised in space Heating) through ranges and open and closed stoves (with and without back boilers) to small industrial furnaces and boilers of many types with efficiencies of up to about 40% with specialist stoves operating under certain conditions at higher efficiencies.

Forestry and Forestry Residue Fuels

Commercial technologies in this category are almost entirely conventional combustion technologies and steam turbine/alternator plant.

Dry Agricultural Residues

These residues - particularly straw, are already used on a small scale as a source of heat for drying and for Heating dairies and glasshouses using conventional combustion technologies. Larger scale systems are uncommon in Europe at present.

Municipal Solid Wastes (MSW)

MSW can be treated by incineration (with or without heat recovery), disposal for landfill or combustion for electricity or CHP or anaerobic digestion.

In recent years, incineration alone has become a less preferred option because of the costs of emissions control and electricity generation or CHP now provides a much more commercially attractive option.

Conventional combustion of MSW using mass burn systems developed specifically for mixed and variable fuels, is commercial but the problems and costs of emissions control and monitoring are prompting the search for lower cost technologies that will be cleaner and more efficient. These include increasing levels of waste sorting and recycling, pelletising and gasification to improve the fuel quality and to assist efficient and clean combustion. Anaerobic digestion for MSW is also under development. Already anaerobic digesters are available at demonstration level for the conversion of MSW - the range of commercial technologies currently available is limited, however.

Landfill Gas

In some locations, shortages of and opposition to landfill sites, pollutant emissions and groundwater contamination are restricting the landfill saleable heat, power or CHP option to control and utilization of the gas in existing sites.

Here the preferred means of energy exploitation are:
- on site use;
- sale to third party;
- electricity generation and sale to the grid.

Other Wet Organic Wastes (WOW)

Anaerobic digesters are available for the conversion of WOW and gasification of digested sewage sludge is becoming more common. At present the range of commercial technologies is, however, quite limited.

Combustible Industrial Wastes

With combustible Industrial Wastes, there are many products and many processes. The general lack of cost-effective an non-polluting combustion systems for wastes (particularly into power at sizes below about 2MWe) has severely restricted the utilization of industrial wastes on site and the general tendency is perhaps therefore, towards off-site centralised incineration - sometimes with, sometimes without steam and power generation - or disposal to landfill.

Ethanol

Production of ethanol from the fermentation of crops high in sugars and/or starches is possible but in Europe so far, the economic and financial conditions are not yet favourable to commercial production. Production of ethanol from lignocellulosic plants such as wood is also possible but commercial systems are still under development - a new route via synthesis gas is also under development.

(ii) General Developments

At a general level, work is underway on the development of cleaner and more efficient and more cost effective Biomass Systems.

At the Crop Production and Pre-Processing Stage, improvements include:

- developments in species selection and cloning, and in the technical management of existing and emerging crops - which should lead to substantial increases in crop yields;

- environmental studies of agricultural systems - in particular with regard to reducing potential negative impacts (soil and water pollution, etc.) of energy crops at local and general levels;

- the development of lower cost wood chippers and other comminution methods - which should greatly enhance harvesting and pre-processing activities;

- cheaper and more efficient drying methods and methods for increasing the degree to which dry combustible materials in mixed fuels such as MSW can be separated out from recycleable materials such as plastics.

At the first stage of Conversion, System improvements include:

-   improved combustion systems (often based on
    fluidised beds to increase efficiency and reduce
    emissions);

-   improved gasification systems (to replace direct
    combustion with more efficient, cleaner and more
    controllable systems);

-   improved gas cleaning and cooling systems
    (to allow producer gas to be used in IC engines);

-   improved anaerobic digestion systems and
    construction methods to reduce costs
    and improve productivity and to allow their use
    with other biomass fuels such as MSW;

-   optimized pyrolytic oil properties, especially
    due to stabilization after pyrolysis or upgrading
    (shown to be technically and economically feasible
    by using a catalytic treatment with hydrogen).

At the Final Stage of Conversion, developments in technologies allowing cost effective solid biomass conversion at smaller scales (less than 2-5MWe) than is generally possible at the moment include:

-   changes to IC engines to allow the use of producer gas;

-   modernization of steam engines to provide lower cost,
    more effective prime movers for small scale CHP systems
    where process steam is needed;

-   the development of stirling engines to provide low cost,
    high efficiency conversion (up to 45% should be possible
    using gasification of dry fuels) of all biomass fuels, refined
    or unrefined at scales ranging from 10kWe to 5 MWe;

-   modified conventional gas turbines that will operate
    effectively on producer gas at 0.5 - 2 MWe;

-   ceramic gas turbines of 100kWe to 1MWe to use
    powdered biomass fuels.

Again, at the Final Stage of Conversion, there have been a number of developments in technologies allowing cost effective solid biomass conversion at the larger scale of up about 50 MWe for systems where large scale biomass production is viable. These include:

- combined cycle gas and steam turbines which are available now for distillate fuels and

- steam injected gas turbines and intercooled, steam injected gas turbines for gasification in systems of about 40-50 MWe.

Potential conversion efficiencies for these systems are estimated at 40-47%. The existing commercial maximum efficiency is about 25%.

5.1.2 Development Prospects

Tables 3 and 4 summarize the principal current technological constraints and opportunities for Biomass, based on the 1990 U.S. Interlaboratory White Paper SERI/TP-260-3674 DE90000322 on the Potential of Renewable Energy.

(a)   Biomass for Electricity Generation:

Table 3. The Principal Current Technological Constraints and Opportunities for Biomass for Electricity Generation.

Issue	Constraint	Opportunity
Conversion Efficiency	Low efficiency combustion	R,D&D on new designs
Market Readiness	Gasification, etc., not yet developed	More R,D&D on newer technologies
Market Compatibility	Uncertainties in energy crop production	Demonstration

(b)   *Biofuels:*

Table 4.. The Principal Current Technological Constraints and Opportunities for Biofuels.

Issue	Constraint	Opportunity
Resource Access	Insufficient variety of high yielding production systems	Genetic, biotechnology R,D&D, productivity demonstrations, harvesting and handling R,D&D
Conversion Efficiency	Better conversion yields needed	Biotechnical R,D&D
Reliability/ Lifetime	Proof of bulk process	System demonstrations at bulk scale

The issues identified in Tables 3 and 4 are discussed in detail in Section 6.

5.2   Economic Review of Biomass Energy

The Review of Present and Predicted Future Performance and Costsof Biomass Energy in Europe presented here, is based on analyses by Grassi, 1992 and by Grassi, Trebbi & Pike et al, in 1992, of those biomass technologies for fuels and electricity that are of greatest interest in the Community at present.

5.2.1 Electricity

5.2.1.1 Cost Determinants

The costs of electricity production from biomass are determined by:

- specific investment;
- capacity;
- efficiency;
- biomass / fuel costs;
- credits where applicable;
- operation;
- lifetime.

(i) Specific Investment

Specific Investments range from around:

-       2000 ECU/kW in the case of direct combustion of solid biomass in power stations, down to
-       1200 ECU/kW in the case of thermal power stations fuelled by bio-crude-oil or water charcoal slurry, and
-       between 330 and 2000 ECU/kW in the case of electricity production by advanced technologies.

(ii) Capacity

Capacities vary from around:

-       2 - 30 MW in the case of direct combustion of solid biomass in power stations to
-       between 100 kW and 50 MW in the case of electricity production by advanced technologies.

(iii) Efficiency

Efficiencies range from around:

-       20% in the case of direct combustion of solid biomass in power stations to

- 35% (power generation) in the case of thermal power stations fuelled by bio-crude-oil or water charcoal slurry, and ..
- about 40% in the case of electricity production by advanced technologies.

(iv) Biomass / Fuel Costs

Biomass / Fuel Costs range from around:

- biomass costs of 50 ECU/d.tonne, in the case of direct combustion of solid biomass in power stations to
- biomass costs of 50 ECU/d.tonne and bio-crude-oil production costs of 210 ECU/TOE, in the case of thermal power stations fuelled by bio-crude-oil or water charcoal slurry, and
- biomass costs of 50 ECU/d.tonne and possible fuel costs (80 μ powder) of 150 ECU/TOE , up-graded bio-crude-oil of about 300 ECU/TOE, bio-crude-oil production costs of about 170 ECU/TOE and expected partially up-graded bio-fuel costs of about 300 ECU/TOE, in the case of electricity production by advanced technologies.

(v) Credits where applicable

Credits may include, for example, an estimated 40 - 84 ECU/TOE (depending on the sulphur content being removed) desulphurization credit in the case of thermal power stations fuelled by bio-crude-oil or water charcoal slurry.

(vi) Operation

In the case of electricity production by gasification of biomass and utilization of aero-engine derived gas turbine/steam turbine combined cycles, for example, economic analysis would assume operation in the region of 7000 hr/year.

(vii) Lifetime

An economic lifetime of 20 years is generally used for the analysis of generation costs.

5.2.1.2 Overall Electricity Production Costs

The foregoing costings/estimates result in overall electricity production costs of:

- 0.11 ECU/kWh in the case of direct combustion of solid biomass in power stations, to

- 0.06 ECU/kWh (base-load) in the case of thermal power stations fuelled by bio-crude-oil or water charcoal slurry, and

- between 0.05 and 0.08 ECU/kWh in the case of electricity production by advanced technologies.

5.2.2 Fuels

5.2.2.1 The Cost of Biomass Pyrolysis Liquids Production and Upgrading for Small Scale Systems.

For small scale systems, the cost of biomass pyrolysis liquids production and upgrading is determined by:

- biomass costs;
- investment costs;
- conversion costs;
- credits.

(Note that the analysis presented here is based largely on a similar analysis by Grassi, Trebbi & Pike et al (1992), and assumes that properly prepared feedstock is delivered to the site and used immediately with no reception, storage or pretreatment operations required - and just minimal handling. This, therefore, represents the most optimistic cost scenario and is restricted to small scale operations of not more than 2 tonnes per hour.)

(i) Biomass Costs

At a yield of 70% by weight, biomass at 50 ECU per tonne can be assumed, giving a cost of 71.4 ECU per tonne of bio-oil at an output of 1 tonne per hour.

(ii) Investment Costs

Investment costs for small scale systems are in the region of 14.3 ECU per tonne of bio-oil at an output of 1 tonne per hour.

(iii) Conversion Costs

Conversion costs for small scale systems are approximately 32.2 ECU per tonne of bio-oil at an output of 1 tonne per hour.

(iv) Credits

The analysis assumes that the following credits are applicable:

- a premium for low sulphur emissions;

- socio-economic and environmental credits associated with labour;

- exemption from carbon tax and subvention from tax revenues to support solutions to the carbon dioxide emissions problem.

5.2.2.2 Production Costs for Hydrotreating and Refining Pyrolysis Oil from Biomass in a Large Scale, Integrated Conversion Process

Overall Production Costs for Hydrotreating and Refining Pyrolysis Oil from Biomass in a Large Scale, Integrated Conversion Process are determined by the Costs of:

- feedstock, e.g. dry, ash-free wood feedstock;
- pyrolysis oil output (including water);
- hydrotreated bio-oil;
- partially upgraded bio-oil;
- refined hydrotreated bio-oil.

(Note: Basis - 1000 tonnes a day of dry, ash-free wood feedstock delivered to the factory gate at a price of 50 ECU/tonne)

(i) Feedstock, e.g. dry, ash-free wood Feedstock

Dry, ash-free wood Feedstock yields of 100% by weight may be assumed:- this corresponds to a cost of 50 ECU per tonne of product and 108 ECU per tonne of oil equivalent.

(ii) Pyrolysis oil Output (including water)

Pyrolysis oil Output (including water) yields of 70% by weight may be assumed:- this corresponds to a cost of 109 ECU per tonne of product and 246 ECU per tonne of oil equivalent.

(iii) Hydrotreated Bio-oil

Hydrotreated Bio-oil yields of 26% by weight may be assumed:- this corresponds to a cost of 384 ECU per tonne of product and 393 ECU per tonne of oil equivalent.

(iv) Partially Upgraded Bio-oil

Partially Upgraded Bio-oil yields of 48% by weight may be assumed:- corresponding costs were not available at the time of writing.

(v) Refined Hydrotreated Bio-oil

Refined Hydrotreated Bio-oil yields of 23% by weight may be assumed:- this corresponds to a cost of 440 ECU per tonne of product and 430 ECU per tonne of oil equivalent.

5.3 Environmental Impacts and Public Acceptability of Biomass Energy in the European Context

The principal environmental impact / public acceptability issues relating to biomass energy in Europe are:

- marginal land use / impoverishment of soil nutrients and erosion resulting from the removal of organic residues from the field;
- airborne and waterborne pollutants;
- visual impacts of biomass plantations and power generating stations;
- health and safety.

(i)   Marginal Land Use / Impoverishment of Soil Nutrients and Erosion resulting from the Removal of Organic Residues from the Field

While biomass could have positive benefits with regard to marginal land use, another land use issue of particular concern, is that of the potential removal of organic residues from the field, leading to impoverishment of soil nutrients and erosion.

A number of studies cited in Giolitti and Jez, et al (1991) have shown that the quantity of biomass which can be removed without significantly affecting the carbon cycle varies from 20% to 50% of the quantity available.

The removal of biomass beyond these limits can lead to further aggravation of those erosion problems already caused by the use of intensive methods of farming.

Erosion in turn deprives the soil of inorganic (N, P, K, Ca) and organic nutrients concentrated in the surface layers. It also reduces the water available to plants, restricts the space for root penetration and reduces the filtering capacity of the soil - thus leading to loss of productivity and eventually desertification.

The ploughing-in of crop residues may help to redress the balance to an extent - but, as Giolitti and Jez et al (1991) observe, even this operation (the efficacy of which depends on many factors: even distribution, ploughing-in to a particular depth, and so on), does not guarantee the maintenance of soil fertility.

These problems must be considered whether it is a case of removing just residues, or an entire optimized energy crop.

Rational management of resources with appropriate farming methods are necessary in order to maintain high levels of productivity and land conservation.

Giolitti and Jez et al (1991) suggest that:

- adequate vegetational cover or other natural barriers must be maintained in the field and surrounding area;

- ploughing depth must be reduced and suitable machinery used, especially on hillside parcels;

- crop rotation must be practised;

- organic fertilizers must be provided.

(ii)   Airborne and Waterborne Pollutants

Both biological and thermal conversion systems produce gaseous emissions. As Cavanagh (1991) remarks:

- airborne emissions from thermal systems include particulates (fly ash), hydrocarbons and other organic products of partial combustion and oxides of nitrogen (NO_x);

- gasification, pyrolysis and catalytic systems may also produce such products, the extent depending on the design of the system;

- combustion in a turbine or engine also produces a similar range, although particulate emissions will generally be low as pre-clean-up systems will have been installed in order to reduce engine damage;

- all thermal systems produce carbon dioxide but as this CO_2 is produced in a closed cycle, it does not add to the overall atmospheric CO_2 concentration;

- anaerobic digesters produce methane;

- where biologically derived fuels are used in engines combustion products may differ from those from direct combustion of the biomass, with increased release of aldehydes;

- in combustion, the formation of particulate emissions is affected by the fuel feed rate, the level of fines, the amount of excess air used and the distribution of the combustion air;

- gas streams from pyrolysis or gasification may contain polyaromatic hydrocarbons such as fluoranthen, chrysen, phenanthrene, benzanthracene, pyrene and benzopyrene. ... these products should, however, be destroyed by combustion where the gases are used either in steam boilers or internal combustion engines;

- with anaerobic digestion, the main problems arise from the generation of toxic hydrogen sulphide ... gaseous emissions from digester, distilleries and thermal conversion plant often emit odours which are unavoidable - though they can be minimized by good initial design ... at the local level, product gases from most conversion and end use systems can be lethal and contribute to explosive mixtures when contained in restricted spaces - thus, adequate ventilation, proper fluesystems, maintenance to avoid leaks and good plant and building design as well as established safety procedures are essential to prevent injury or death.

The most toxic biomass liquid effluent is the carcinogenic liquid fraction from pyrolysis plants - the safe disposal of which poses a particular problem for small scale rural systems. In large scale plants, liquids can, however, be injected back into the system - or treated by aerobic biological systems.

The principal solid waste produced from biomass generation systems is fuel ash. This and other solids may be treated or returned to the land as fertilizer.

Finally, noise impacts in biomass based power generation can be reduced by housing generation plant in suitable buildings with the generators below ground level. It is more difficult to reduce the noise and vibrations associated with the transport, handling, milling and conveying biomass, however, and local noise is largely unavoidable.

(iii) Visual Impacts of Biomass Plantations and Power Generating Stations

Large biomass plantations will alter the landscape - and large power generating stations can cover several hectares with large buildings, stacks, conveyors and silos or piles of raw materials.

Environmental impact analysis, planning and regulatory compliance can, however, minimize negative and maximize positive effects.

(iv) Health and Safety

Potential occupational risks include:

- risks associated with exposure to pesticides and herbicides in wood fuel plantations;
- risk of accidents from tree harvesting and mechanical equipment;
- risk of accidents in wood fired power plants.

Potential public risks include:

- health risks from possible surface and groundwater
  contamination from pesticides, herbicides and fertilizers;
- risk of accidental injury from road transportation;
- risks of cancer associated with exposure to polycyclic
  organic matter (wood has a much lower air polluting
  SO_2 emission level than does coal or oil - but higher
  levels of particle emissions - especially polycyclic
  organic matter, exposure to which has been linked to
  cancer).

Increasing interest in biomass may be attributed to a growing awareness of its various economic, environmental and political advantages over other fuels.

If biomass is grown sustainably, the amount of carbon dioxide released in its burning is balanced by the amount taken in during photosynthesis. Thus the utilization of biomass for energy allows a significant amount of renewable fuels to be consumed without increasing the CO_2 content in the atmosphere and while allowing partial substitution of conventional fuels. Furthermore, bio-fuels contain minimal sulphur - thus minimizing SO_2 emissions.

Many of the potential environmental impacts associated with using energy crops are common to other agricultural activities. Recycling of compost derived from biomass as a soil conditioner reduces soil deterioration and pollution of ground water.

If operated correctly, modern bio-energy technologies and bio-fuels are relatively benign and produce very little pollution. They have negligible deleterious effects on air, land or water when using state-of-the-art environmental control technologies.

5.4 The Commercialization of Biomass Energy in Europe

Current biomass usage in the Community totals 50 Mtoe and is divided almost equally between utilization for industrial products (30 Mtoe) and utilization for energy (20 Mtoe) (Saarbrucken, 1988 and Hall and Rosillo-Calle, 1990).

EC Targets for advanced biomass technology implementation are (after Grassi and Bridgwater, 1990):

-	demonstration of:	
	- short rotation forestry:	1994
	- advanced combustion for power:	1995
	- advanced pyrolysis:	1995
	- pulp for paper from sweet sorghum:	1995
	- compost:	1996
	- sweet sorghum production:	1996
	- char slurry fired boilers, kilns etc.:	1996
	- gas turbine fired on bio-oil and char slurry:	1997
	- advanced gasification:	1998
	- hydrocarbon synthesis:	2000
-	commercialization of:	
	- advanced combustion for power:	1996
	- compost:	1998
	- advanced pyrolysis:	1998
	- short rotation forestry:	2000
	- sweet sorghum production:	2000
	- ethanol for fuel:	2000
	- advanced gasification:	2003
	- methanol synthesis:	2003
	- hydrocarbon synthesis:	2005
	- hydrogen and ammonia production:	2010

There are three commercial energy channels open to biomass. These are:

- solid fuels (underboiler/kiln);
- liquid fuels (automotive);
- electricity.

Solid Fuels:

With regard to solid fuels, Grassi et al (1992) observe that:

> - solid biomass fuels have a low energy density - particularly
> in the raw state, so commercial opportunities in kiln or underboiler
> applications are limited by energy market economics

(transport costs in particular) to locations close to the point of biomass production;

- niche opportunities often involve specific local factors such as the need to dispose of process residues - or the availability of wastes;

- sometimes investment grants are available for specialised combustion systems for which conversion costs can be high;

- extending the application of solid biomass fules to meet local underbiler requiremetns could stimulate further upgrading of fuel oil into the transportation sector which is less easily penetrated by renewables.

Liquid Fuels:

In the early 1980s, when oil prices peaked at around $(1990)60/barrel, the automotive market was seen as the logical target for biomass conversion processes. Today, the alcohols (ethanol and methanol) are perceived to be the most important near term renewable transportation fuels (ethanol is already used widely in Brazil and also in the US and other countries such as Zimbabwe).

In Europe, at present, the production of ethanol and methanol as an ingredient for motor fuels is considered feasible in the medium to long term. The production of pyrolytic oil or bio-oil is also considered promising in the medium term and its integration into conventional energy systems would appear, at present, to present no major difficulties.

Electricity:

At present, electricity is the most attractive, large, near-term market for biomass resources in the EC. The main technologies which can be envisaged for the production of electricity from biomass resources were identified by Grassi, in November 1991, at the 'First European Forum on Electricity Production from Biomass and Solid Wastes by Advanced Technologies', as:

- direct combustion of solid biomass in power stations - already commercial with electricity production costs of about 0.11 ECU/kWh (assuming biomass costs of 50 ECU/d.tonne; specific system investment costs of 2000 ECU/kW; a 2-30MW capacity; 20% efficiency);

- thermal power stations fuelled by bio-crude-oil or water charcoal slurry - it has already been demonstrated that bio-crude-oil, obtained by pyrolysis, can be fired directly into boilers .. the commercial production of bio-crude-oil should be feasible around 1995 with electricity (base-load) production costs of about 0.06 ECU/kWh (assuming biomass costs of 50 ECU/d.tonne; 35% efficiency;

bio-crude-oil production costs of 210 ECU/TOE;
an estimated 40-84 ECU/TOE credit for desulphurization -
depending on the sulphur content being removed; and
investment costs of 1200 ECU/kW);

- electricity production by advanced technologies:

  - decentralised electricity production by firing small
    biomass powder in ceramic gas-turbines - with
    commercialization around 1996 and electricity costs
    (16hr/day) of about 0.05 ECU/kWh (at a biomass
    cost of 50 ECU/TOE; 20 year life;
    fuel cost (80 m powder) of 150 ECU/TOE;
    target efficiency of 40%; capacity of 100-500 kWe;
    specific investment (full system) of 330 ECU/kW and
    a penetration rate of 1000-5000 MW/y);

  - electricity production by aero-engine derived gas
    turbine-steam turbine combined cycles fuelled by
    up-graded bio-crude oil - commercialization around
    1996 with electricity production costs of about 0.07
    ECU/kWh (base-power 7000 hr/y) / 0.08 ECU/kWh
    peak-power 4200 hr/y) (this assumes biomass cost of
    50 ECU/d.tonne; 0.3-50 MW capacity; bio-crude-oil
    production cost of about 170 ECU/TOE; expected
    partially upgraded bio-fuel cost of about
    300 ECU/TOE; 500 ECU/kW investment for
    biomass conversion and upgrading; 20 year life);

  - electricity production by gasification of biomass
    and utilization of aero-engine derived gas
    turbine/steam turbine combined cycle with
    electricity production costs of around
    0.067 ECU/kWh (this assumes biomass cost of
    50 ECU/d.tonne; capacity of 1-30 MW (lower
    than for liquid fuel systems); system efficiency of
    (0.80 (gasifier) x 0.50 (power generation)) = 40%;
    specific investment (gasifier + power generation) of
    2000 ECU/kW; 20 year life; and 7000 hr/year
    operation).

Thus, the most advanced and promising concepts for power generation from biomass, require gas turbine / steam turbine combined systems - having the characteristics and advantages of the multi-fuel generators - capable of being fed by a wide range of bio-fuels, including: bio-gas; up-graded bio-crude-oil; ethanol; methanol; slurries and biomass powder.

Finally, it should be noted that biomass can provide electricity at a cost comparable with the true cost of power from conventional sources (that is: when account is taken of the costs of pollution, waste-disposal, etc.);

At a more general level, biomass for fuels or electricity, can:

- offer a new market for EC agriculture - the potential of biomass energy plantations is of great significance for the future ... but whether this potential can be realised is largely dependant on its development in the future as a non-food production option within the Common Agricultural Policy;

- provide new opportunities for rural development;

- make significant contributions to the improvement of global and local environments.

The future potential of biomass energy markets as a whole, will, however, depend on:

- in the case of energy cropping:

  (a) the amount of land which will be made available for energy farming and

  (b) biomass crop productivities;

- source-to-fuel conversion and upgrading technologies;

- the development of engines capable of running efficiently on bio-fuels, producing acceptable emissions, available at sizes compatible with power demands and having low capital and running costs.

It is also essential that institutional factors such as tariff barriers or differential taxation do not obstruct the development of biomass energy markets.

The theoretical potential biomass share of Community energy consumption by the year 2000 has been estimated to be in the region of 4% - the potential which is likely to be economically viable, however is about 3%.

6 Conclusions: Prospects for the Development of Renewable Energy in Europe to the year 2000.

6.1 Overview of the Current Status, in Europe, of the four specific Renewable Energy Technologies under review

Table 5 presents a summary of the current degree of technical maturity and market readiness of the four Renewable Energies under review - based on information supplied in a 1992 report produced by UNSEGED (the UN Solar Energy Group on Environment and Development).

Table 5. The Current Degree of Technical Maturity and Market Readiness of Wind, Solar Heating & Cooling and Daylighting, Photovoltaics and Biomass.

TECHNOLOGY	RATING				
	1	2	3	4	5
Wind					
- Installations of up to 400 kW	✓	✓	✓		
- Larger installations				✓	
Solar Heating & Cooling and Daylighting					
- Active systems	✓		✓		
- Passive systems	✓	✓	✓		
Photovoltaics					
- PV-application-technologies (e.g. pumps, lights)	✓	✓		✓	
- PV-electricity (a-Si)				✓	✓
- PV-electricity (other materials)				✓	✓
Biomass					
- Combustion	✓	✓		✓	
- Gasification	✓	✓		✓	
- Bio-alcohol	✓	✓			
- Vegetable Oil	✓	✓			
- Other Energy Crops				✓	✓

Key:

1: technically mature for market (possibly with need of optimization);
2: more application experience needed (e.g. lifetime of equipment);
3: R&D needed for alternative solutions;
4: development needed;
5: research needed.

Average estimated Electricity Production Costs (ECU/kWh) range between:

-   0.05 ECU/kWh now and 0.03 ECU/kWh by the year
    2000, for Wind;
-   0.6 ECU/kWh now and 0.3 ECU/kWh by the year
    2000, for Photovoltaics;
-   0.1 ECU/kWh now and 0.05 ECU/kWh by the year
    2000, for Biomass.

Fuels costs (from Biomass) are estimated at:

-   solid:
    100 ECU/tce for cellulosic material (compared
    to 130 ECU per tonne of domestic coal);

-   liquid:
    60 ECU/barrel of bio-alcohol from wheat;
    40 ECU/barrel of pyrolysis oil and
                    refined pyrolysis oil;
    (compared to approx. 20 ECU/barrel for petrol).

6.2 Conservative Estimates on Share of Community Energy Consumption by the End of the Century

The conservative estimates which follow are based on figures supplied in 1992, by DGXVII.

An increase in R,D&D and market stimulation activities as outlined throughout this Report, would, of course, serve to hasten the deployment of Renewables across the Community.

Wind

The estimated share of Community energy consumption by the year 2000 for wind energy under present 'natural growth' conditions, is about 1% of electricity demand.

Solar Heating & Cooling and Daylighting

The estimated share of Community energy consumption by the year 2000 for solar heating & cooling and daylighting technologies - active and passive, under present 'natural growth' conditions, is 1-2%.

Photovoltaics

Unless the current annual rate of photovoltaic installations increases substantially, the estimated photovoltaics share of the Community's energy consumption by the year 2000 will be marginal under present 'natural growth' conditions.

Biomass

The theoretical potential biomass share of Community energy consumption by the year 2000 has been estimated to be in the region of 4% - the potential which is likely to be economically viable under present 'natural growth' conditions, however, is about 3%.

6.3   Factors affecting Further Development

6.3.1 Development of Renewable Energy Technologies

6.3.1.1 Setting Research, Development and Demonstration Priorities

(i)   Wind:

Notwithstanding the fact that mature wind turbine designs are demonstrating good performance and reliability, there is still scope for improvement.

Factors limiting the reliability, performance and economics of wind energy run the whole gamut from gearbox, generator and brake system problems, to blade, hub, shaft and tower fatigue failures; tower and rotor aero-elastic instability induced failures; transmission problems and bearing and control system failures - and, of course, project planning factors (e.g. permits, etc.).

Significantly:

> - EUROWIN's failure analysis of control system, yaw system, rotor blades, generator, grid, mechanical brakes, gearbox and other failures, indicate that, at present, the control and yaw systems are the most problematic components;

> - poor performance levels have also been attributed to blade soiling, poor siting (that is: in areas of low wind speed and/or high turbulence), wake effects and poor or non-existent maintenance;

> - questions of optimum size of wind turbines, vertical versus horizontal axis, lifetime prediction and so on remain to be finally resolved - though they are not vital for short term developments;

- there is still scope for cost reduction which can be achieved at least in part by further improvements in the technology;

- in order to fulfil ambitious expansion programme requirements, the utilities need high-performance and highly reliable MW-size machines;

- finally, consideration of offshore sites, significantly increases the number of suitable locations which may be selected for future wind power plants (the annual output for an offshore windfarm can be up to 40% higher than that of a landbased windfarm if turbines designed for wind conditions on land are used - turbines modified to wind conditions at sea could further increase this output).

Continued Research and Development is therefore required in the following Areas:

- Resource and Siting, particularly with regard to:

  - the improvement and simplification of methodologies for resource characteristics information gathering and prediction need to be improved and simplified;

  - detailed topology analysis which is required for site selection, particularly in the case of multi-machine installation proposals.

- Turbine Design and Components, particularly with regard to:

  - improvements in the areas of electricalconversion systems (variable speeds), control mechanisms, yaw response and control, fatigue loadings, tower design and blade shape and mounting design;

  - aerodynamics and mechanics with a view to reducing dynamic loads;

  - materials and manufacturing methodology;

  - materials, manufacturing and componentevaluation activities with a view to greater reliability, greater safety and lower costs (application of probabilistic reliability analysis).

- Installation, Operation and Maintenance, particularly with regard to:

  - applications other than grid-connected systems - research into system operation is required in order to consolidate knowledge gained thusfar;

- fault detection, prediction and propagation - including the development of new methods such as Non-Destructive Testing (NDT) which can be applied to the large components and structures involved without causing significant disruption to power generation.

- Energy, Applications and Environmental Integration, particularly with regard to:

- possible localised level regional integration problems with power transfer from areas of high wind intensities to areas of high demand;

- applications other than onshore electricity generation (offshore systems, systems sharing land use with agriculture, direct mechanical power for water pumping or milling) ... particularly with regard to component development for design integration and optimization;

- further work on Standards, Regulations and Communication to address design and testing issues and environmental integration issues such as visual impact, land use, noise, impact on wildlife and natural habitat, telecommunications interference and safety... there is also a need for harmonized certification procedures and rules.

(ii)   *Solar Heating & Cooling and Daylighting:*

Work must continue in the following broad areas:-

- Collection;
- Conservation: Thermal Insulation and Controls;
- Waste Heat Recovery;
- Thermal Energy Storage and Distribution;
- Comfort Analysis;
- Micro-Climate;
- Heat Attenuation;
- Ventilation and Air Circulation;
- Natural Heating/Cooling/Lighting Techniques;
- Monitoring in Heating, Cooling and Lighting;
- Control Strategies and Systems for heating, cooling and lighting;
- Performance Criteria, Integration and Standards;
- Design Support and Tools and Model Development.

Of these, the five areas receiving particular attention at present are:

(a)   *Collection - particularly in the area of Active Systems:*

In order to maximize net Solar gain in active systems, research is continuing on the following:

- collector orientation and tilt;
- reflection and reflector installation;
- glazing with high levels of solar radiation transmittance and low levels of thermal transmittance;
- heat retention, the installation of movable insulation (used only at night in a direct system) and transparent insulation materials;
- the overall design of the system to achieve a high absorptance of solar radiation at appropriate times of day and year.

(b)   *Thermal Energy Storage and Distribution:*

The suitability of any particular thermal storage type (primary, secondary and remote) for a particular application, depends on:

- the size and material of the storage;
- the means by which solar heat is charged and released;
- the insulator used.

Research to date has focussed on:

- storage materials;
- storage charge and discharge;
- effects of thermal mass.

Progress in the area of thermal energy storage has been slow over the last few years and has been predominantly in the area of optimizing domestic hot water storage.

Work continues, however, on its improvement, particularly in incorporating phase change materials in building components and on transparent insulating materials and there has been significant improvement in the performance of the window in relation to both heat and light.

An appreciation of the dynamic potential of glazed apertures is generating various louvres, shutters and shading devices (manually and automatically controlled), and interest is growing in 'smart' glazing materials including thermochromic or electrochromic and reflective or transmissive holographic window coatings. Improved fenestration materials increase the need for 'smart' building controls which integrate the various Systems.

There is also much interest in seasonal storage at present - research concentrating on lining and insulation of the store.

Clearly the most efficient way to distribute passive solar energy is to design the layout of the rooms such that the solar energy is collected and stored in or adjacent to the location of its intended use:- this is known as Thermal Zoning.

Remote Storage Systems, however, require mechanisms for distributing heat to those areas requiring it. In existing buildings, central air heating systems may already regulate the distribution by means of the re-circulation of air - but designing a building such that distribution takes place exclusively and effectively by means of natural air flows requires care.

(c)   Materials:

New materials will play an increasingly important role in the future. Examples of new types of materials on the market are heat reflective coatings, shading devices, phase change materials for low-temperature storage of heat, holographic panels for filtering and redirecting daylight, electrochromic glasses and transparent insulation.

Future research will focus on assessment of material properties, lifetime predictions and energy-saving potential.

(d)   Performance Criteria, Integration and Standards:

The definition and introduction of an energy performance standard (and reference energy audit procedures and tools) for buildings in which all major energy-saving measures are incorporated would serve as a major encouragement in this area.

(e)   Design Support and Tools and Model Development:

Considerable challenges exist in developing services in support of the building design professions so that integrated design tools and building performance evaluation methods as well as up-to-date technical information is readily available to the European construction industry and the individual practitioner.

Increasingly, computer models are making it possible to optimize designs with regard to energy consumption and thermal comfort - and improved simulation models will enable the development of better evaluation tools.

(iii) Photovoltaics:

A number of technical issues remain outstanding for all PV technologies as outlined in Section 4.1.

Future developments in PV systems will aim at achieving high module efficiencies, low cost cell and module production processes, low cost support structures and high efficiency power conditioning equipment.

(a) Module Efficiency

Module efficiency depends on both the semiconductor cells within the module and the encapsulation of these cells. Currently, the highest efficiencies of crystalline cells in production are 17% with corresponding module efficiencies of 15%. However laboratory efficiencies of 23% have already been achieved and within the next few years modules with efficiencies of 19%-20% will be developed. The improvement in module efficiencies will be brought about by:

- developing new cell structures
  e.g. the LGBG (laser grooved buried grid) cells;
- reducing surface reflectance;
- development of tandem a-Si/c-Si cells with projected
  efficiencies of 25%;
- increasing the module packing factor;
- reduction of module mismatch and wiring losses.

With regard to thin film technologies substantial development work is required to reduce light induced degradation and in achieving stable efficiencies between 10% to 15%. The developments in these areas include:

- multi junction structures;
- researching new compound semiconductors
  e.g. CIS and Cadmium Telluride.

In addition to improving module efficiencies, research must also continue on extending the module lifetimes to 30 years. This will involve improving the resistance of the module materials (glass metal and plastics) to corrosion.

(b) Cell and Module Production

The principal issues in cell and module production processes are:

- lowering cost/Wp;
- expanding the manufacturing capacity.

The main factors affecting cell cost are the feedstock material, crystallization and the wafer sawing. Module costs are determined by the encapsulation materials, cell soldering, cell encapsulation and module framing.

The key development areas for the optimisation of module manufacturing are:

- reduction of the wafer thickness;
- replacing diamond saws with multi-ingot wire saws;
- processes which eliminate the sawing step;
- increasing automation of the manufacturing processes;
- direct sheet/ribbon production techniques;
- reducing cell breakages on the production line;
- large cell (>400cm^2) and module(>1kWp) production;
- eliminating the metallic frames around modules;
- lower cost encapsulating materials and procedures.

(c)   *Module Support Structures*

Module mounting structures account for over 20% of the total cost/Wp and are directly proportional to the area occupied by the modules. Hence, as module efficiencies improve support structure costs will fall in proportion. However, specific methods for reducing module mounting costs are currently being developed. These include:

- light weight and stretched cable structures;
- integrating modules into existing structures e.g. roofs and facade cladding. Amorphous Si modules are particularly suited to cladding applications.
- gluing the modules to the metal frame;
- reliable labour saving methods for connecting modules together.

Reductions in module weight/Wp will also provide additional savings in support structures.

(d)   *Power Conversion and Conditioning*

Developments in inverter technologies aim at achieving high efficiency, low cost, low harmonic distortion and high reliability. As the rated powers of inverters continues to increase so also will the commutation losses of the power semiconductors Increasing the switching frequency of the power semiconductors reduces the level of harmonic distortion of the current supplied to the grid but increases the commutation losses. Some of the developments taking place are:

- <u>power switching devices:</u>

the recent advent of IGBTS (Insulated Gate Bipolar Transistors) has meant lower switching losses in new inverter designs up to 50kW; further developments in discrete semiconductor technology will extend this range to 1 MW.

- high frequency switching:

PWM (Pulse width modulated) switching circuits are being developed to replace conventional 6 and 12 pulse firing circuits; for high power applications (>100kW); these new switching techniques will reduce the level of harmonic distortion and physical size of the inverter components, especially the inductors, capacitors and isolating transformers; reduction in the sizes of these components will be maintained by increasing the switching frequencies (>100kHz); this will only be accomplished by developments in power semiconductors and in new magnetic materials, however.

- integrated module inverters:

low powered line commutated and self commutated inverters (100W) have recently been developed which mount on to the rear side of PV modules; the advantages of these new systems are elimination of DC wiring, improved module matching and increased modularity (PV systems are easily extended at very low cost).

- modular power conditioning:

in order to meet the requirements in decentralized applications, from special loads (for example: pumps) to island grids, different storage units for short-, mid- and long- time frames need to be integrated. Developments will aim at making AC-coupled components more easily integratable and more flexible - thereby ensuring high reliability and scope for expansion in each application.

(iv)   Biomass:

At a general level, work is underway on the development of cleaner and more efficient and more cost effective Biomass Systems.

At the Crop Production and Pre-Processing Stage, improvements include:

-    developments in species selection and cloning, and in the technical management of existing and emerging crops - which should lead to substantial increases in crop yields;

-    environmental studies of agricultural systems - in particular with regard to reducing potential negative impacts (soil and water pollution, etc.) of energy crops at local and general levels;

- the development of lower cost wood chippers and other comminution methods - which should greatly enhance harvesting and pre-processing activities;

- cheaper and more efficient drying methods and methods for increasing the degree to which dry combustible materials in mixed fuels such as MSW can be separated out from recycleable materials such as plastics.

At the first stage of Conversion, System improvements include:

- improved combustion systems (often based on fluidised beds to increase efficiency and reduce emissions);

- improved gasification systems (to replace direct combustion with more efficient, cleaner and more controllable systems);

- improved gas cleaning and cooling systems (to allow producer gas to be used in IC engines);

- improved anaerobic digestion systems and construction methods to reduce costs and improve productivity and to allow their use with other biomass fuels such as MSW;

- optimized pyrolytic oil properties, especially due to stabilization after pyrolysis or upgrading (shown to be technically and economically feasible by using a catalytic treatment with hydrogen).

At the Final Stage of Conversion, developments in technologies allowing cost effective solid biomass conversion at smaller scales (less than 2-5MWe) than is generally possible at the moment include:

- changes to IC engines to allow the use of producer gas;

- modernization of steam engines to provide lower cost, more effective prime movers for small scale CHP systems where process steam is needed;

- the development of stirling engines to provide low cost, high efficiency conversion (up to 45% should be possible using gasification of dry fuels) of all biomass fuels, refined or unrefined at scales ranging from 10kWe to 5 MWe;

- modified conventional gas turbines that will operate effectively on producer gas at 0.5 - 2 MWe;

- ceramic gas turbines of 100kWe to 1MWe to use powdered biomass fuels.

Again, at the Final Stage of Conversion, there have been a number of developments in technologies allowing cost effective solid biomass conversion at the larger scale of up about 50 MWe for systems where large scale biomass production is viable. These include:

- combined cycle gas and steam turbines which are available now for distillate fuels and

- steam injected gas turbines and intercooled, steam injected gas turbines for gasification in systems of about 40-50 MWe.

Potential conversion efficiencies for these systems are estimated at 40-47%. The existing commercial maximum efficiency is about 25%.

The principal developments in the next few years will be:

- new sources / improved production techniques;

- increased production of ethanol for the transport sector from high sugar-/starch- crops such as sweet sorghum;

- use of vegetable oils such as rape-seed oil as a diesel substitute;

- availability of various bio-oils (pyrolytic) in substantial amounts and development of bio-oil stabilization and upgrading processes, as well as increased use of (refined) pyrolysis oil - for example, for turbines for peak electricity production;

- commercialization of new lignocellulosic crops, both perennial woody short rotation crops and annual and perennial herbaceous crops;

- more widespread harvesting of forestry residues for energy;

- development of infrastructure for regional integration.

While the technology exists for the second and third, additional effort is required with regard to converting the fifth to electricity (to provide lower costs and higher efficiencies) and to liquid fuels (particularly in the area of hydrolysis).

6.3.1.2 Research, Development and Demonstration Financing

In addition to ongoing national R,D&D programmes within the Community, work on Renewables is also supported by the European Commission - to date, through its JOULE and THERMIE programmes and soon, through ALTENER.

Normally these programmes are heavily oversubscribed:- this may be taken to indicate a requirement for greater commitment and financial backing by the Commission in this area.

In a technology growth area such as Renewables, the required R,D&D budget is normally taken to be about 5% of turnover.

Given the rapid technological advances and resulting substantial decreases in production costs over the last decade, it is clear that if properly supported, research and demonstration activities and the dissemination of information on Renewables could significantly improve related European and world market penetration prospects.

Given:

     (i)   the Community's strategic dependence where energy is concerned - and the continuing insecurity of traditional energy source supplies;

     (ii)   the importance of Renewables for future environmentally acceptable energy supplies;

     (iii) the under-exploited potential of Renewables in Europe and the world today ...

there is ample justification for accelerating the realization of the potential of Renewables through increased R,D&D support.

6.3.1.3 Conclusions - a Strategy for Technological Development

'Technology-push' through increased national and EC level R,D&D support and budget allocation in the areas indicated, can significantly increase the rate of market uptake of Renewables - thus impacting on the success of the Community's various Renewables Industries and European and world markets in the coming decades.

Public funding for R,D&D on Renewables in the EC in 1992 is estimated at 400 Mio ECU. To achieve the desired goal of greater market penetration by the year 2000, this budget should be multiplied by 3.

The need for Community cohesion requires that these funds are spread more evenly over the Member States - today they are concentrated in a few countries only, viz.:

<u>1992:</u>   150 MECU   Germany
             100 MECU   Italy
              50 MECU   UK
              15 MECU   DGXII
                others   Denmark, Netherlands, Spain,
                        Greece, France
           no budgets   Ireland, Portugal, Belgium,
                        Luxembourg.

6.3.2 Development of the European Market for Renewable Energy Technologies

6.3.2.1 Market Situation / Stimulation

As stated in Section 6.2, a conservative estimated share of Community energy consumption by the year 2000 for wind energy is about 1% of electricity demand; for active and passive solar heating & cooling and daylighting technologies: 1-2%; photovoltaics: marginal; biomass: about 3%.

Thus, Renewables will continue to follow a natural growth rate, making a gradually increasing contribution to the EC energy mix over time - driven largely by technology performance improvements and being gradually manoeuved into an increasingly more cost competitive position vis-a-vis conventional sources of energy ... technologies being pushed nearer to competitive thresholds.

Given the many political, strategic, socio-economic and environmental benefits to be accrued from the realization of the currently vastly under-exploited potential of Renewables, however, there is ample justification for accelerating this growth rate.

Growth rate acceleration may be achieved by combining political commitment and goodwill with 'Technology Push Activities' (that is: R,D&D intensification), 'Market-Pull Activities' (that is: installation subsidies, tax reliefs and premium payments) or both - and, of course, promotional activities based on well conceived, well adapted, full-scale, pilot and demonstration projects.

At national and Community level, policy decisions could be taken and market goals defined in order to set a general frame and express the willingness of the decision-makers to foster the development and market introduction of Renewables.

Attention must be given to the fact that a first prerequisite of market introduction is the development of well designed products and processes - industrially and commercially available for easy utilization and integration.

Regulations are needed to make the existing energy networks accessible to new sources of decentralized and Renewable energy.

Renewable power must be allowed reasonable access to the European grid. Countries rich in Renewable resources would then be able to export electricity to those less fortunately endowed but still wishing to use Renewable Energy in order to honour their committments to environmental protection, for example: to reduce CO_2 emissions. This would, in turn, significantly enhance the degree of market penetration of Renewable Energy in Europe.

Financial support in terms of direct subsidies for development, manufacturing and utilization of Renewable energy products as well as tax deductions, premium payments and cost shifting in favour of Renewables through charges for conventional energies must be considered - Renewables need an equitable market.

Norms and standards must be developed to encourage responsible as well as competitive market growth.

Finally, it must be recognized that upstream of market development, pre-normative R&D and scientific and technological promotion activities continue to represent the starting point of improved performance and reliability of products and systems as well as further cost reductions.

In addition, highly visible and easily understandable demonstration projects can encourage both the public acceptance and market penetration of Renewables in Europe.

Consequently, an increase in R,D&D budgets coupled with the setting of ambitious but realistic, highly visible targets for Renewable Energy production and uptake will send important signals to industry and the general public alike.

As an example of what can be done ...

In the late 1970's, more than 50% of the electricity generated in California was provided from oil, while only small amounts of power were produced by geothermal plants - and the solar-electric, wind and biomass facilities were in their infancy.

By 1991, no oil was used to generate electricity and Renewable energy provided about 25% of the electricity in California. This was achieved through the employment of a combination of strategies, programmes and policies, including: 'set-aside' for Renewable energy in allocation of the need for new electrical capacity, financial incentives and research and development programs.

As a result of these programs, Renewable energy has flourished in California, providing about 5,200 MW of electrical capacity (representing about 12% of California's peak load)... wind energy providing about 1,700 MW of effective capacity, geothermal - 2,500, biomass and municipal solid waste projects - more than 1,200 MW and solar-electric plants - almost 400 MW.

The associated socio-eonomic benefits are many: in total, seven-hundred firms are involved in the Renewables effort and, in addition to benefitting from the avoided costs of oil imports, California now enjoys the second lowest CO_2/NO_x emission levels of all the United States.

Within the EC today, the following accelerative moves could reasonably be made in the short term with regard to the four Renewables under review:

(i)   Wind:

A major effort could be made in the peripheral areas of the Community to generate a significant fraction of national electricity by the year 2000 with the aim of promoting economic development in these areas and decreasing energy import levels.

Relevant areas include: Greece, Sicily, Corsica, Portugal, parts of Spain - and, particularly, the West of Ireland and Scotland.

The current annual rate of installation of wind turbines within the Community is 300MW. It would be realistic to expect that this figure could be effectively tripled between now and the year 2000.

An additional (300 x 3 x 8 =) 7,200 MW of wind power by the year 2000, at a capacity factor of 25%, would correspond to almost 2% of the Community's electricity demand of 1,100 TWh.

With regard to furthering the Community's overall aims regarding Renewables, the additional 200% to be integrated over the next eight years would make just a small difference in terms of installed electrical capacity - it would, however, make a substantial difference in terms of wind energy visibly coming on line within the Community.

A concerted effort should, therefore, be made to set up factories/wind-turbines in at least some of these areas - licensing the technology from existing manufacturers in Denmark and Germany, for example.

Economic studies should also be carried out for each region in order to quantify the potential benefits for disadvantaged regions of an initiative like this in terms of wind turbine production, electricity generation, avoided costs of imported oil, and so on.

(ii)   Solar Heating & Cooling and Daylighting:

EC energy consumption for the heating & cooling and daylighting of buildings is substantial (approximately 50% of all primary energy).

The Community's building stock is replaced at a rate of about 2% per annum. This provides an opportunity for the introduction of legislation regarding energy requirements of new buildings - where building would not be permitted unless basic energy criteria were met. This could be preceeded by a period of optional compliance.

Energy auditing could also be introduced for existing buildings - on a voluntary basis at first - with guidelines on energy consumption/m^2/year.

The UK's BREEAM scheme (developed jointly by ECD and the Building Research Establishment) could be used as a basis of both schemes.

Benefits of such schemes would include the following:

- good buildings would be given a high profile;
- building construction, fuel running and maintenance costs would be reduced;
- the productivity of building users would be increased;
- a significant contribution would be made to environmental improvement.

Both of these activities could then be used to gradually frame broader legislation.

Sweden is currently working on reducing its energy consumption levels in buildings from 200 kWh/m2/year to just 50 kWh/m2/year by the year 2000.

If EC Member States were to adopt the more modest and, certainly, realistic goal of introducing mandatory energy criteria for all new builings built after the year 2000 so that energy consumption could be reduced by half the current level, cumulative energy savings of up to 1% per annum could be achieved.

(iii)   Photovoltaics:

If PV R,D&D efforts/budgets were intensified by a factor of 3, subsequent cell efficiency improvements and cost reductions would mean that by the year 2000, there could be substantial increases in the use of PV:

-on roofs - solar tiles:
a target of at least 10% of all new domestic buildings each year, in each Member State would be feasible;

-in building cladding:
a target of at least 10% of all new offices and industrial buildings each year, in each Member State would be feasible (see Hill, 1992);

-in centralized power stations:
there is an urgent need for a number of large PV demonstration plants across Europe - these would demonstrate potentials and accelerate

movement along the learning curve of how to implement such plants efficiently and well - a realistic implementation target would be at least five new PV power stations of at least 5 MWp in Southern Europe, by the year 2000 (the construction of these large PV plants should be based on connecting together a series of e.g. 100 kWp plant building blocks ... therefore, research should be directed towards the optimization and cost reduction of these modular units).

Higher production volumes would in turn mean further cost reduction and even greater deployment.

Because of the particular attractiveness of PV with regard to the environment, economic growth, North-South co-operation and vast potential markets in developing countries, it is desirable to catalyze the learning process in PV development and production - now.

(iv)   Biomass:

With regard to energy crops, the potential land which will become available for energy industry farming is increased by the changes in land utilization to reduce surplus production of cereals, sugar, wine, milk and beef. This is forecast to be approximately 20m ha by the year 2000.

Estimated production volumes are in the region of 6 tonnes of oil per ha/year.

These figures (20 ha x 6 tonnes) correspond to about 20% of the total energy consumption of Europe.

A more modest and certainly an attainable goal would be the adoption of actions across the Community, following those of France, aimed at biomass substitution of at least 5% of imported hydrocarbons by the year 2000.

With regard to electricity generation from biomass, a realistic target would be that by the year 2000, at least 1% of each Member State's electricity requirements be generated from biomass sources.

In this regard, there is an urgent need for a number of good biomass demonstration plants across Europe - to demonstrate potentials and accelerate movement along the learning curve of how to implement such plants efficiently and well - a realistic implementation target would be at least twenty small (1-5 MW - for islands and developing countries) and five large (50 MW maximum - to act as a prototypes for local generating stations in Europe) demonstration plants by the year 2000.

6.3.2.2 Conclusions - a Strategy for Market Development

'Market-pull' through increased national and EC level market stimulation in the areas indicated, can also significantly increase the rate of market uptake of Renewables - again, impacting on the success of the Community's various Renewables Industries and European and world markets in the coming decades.

6.4   Final Conclusions

Again, given:

(i)   the Community's strategic dependence where energy is concerned - and the continuing insecurity of traditional energy source supplies;

(ii)   the importance of Renewables for future environmentally acceptable energy supplies;

(iii)   the under-exploited potential of Renewables in Europe and the world today;

(iv)   the many socio-economic benefits to be accrued from the pursuit of the Renewables option ...

there is ample justification for accelerating the realization of the potential of Renewables in Europe.

In order to achieve the accelerated realization of Renewables in Europe, there is an urgent need for national governments, the European Parliament and other decision-making bodies to declare their intention to promote Renewable Energies, to define strategies and policies, define overall goals for development and implementation and to decide without delay, on necessary laws and regulations.

Policies should focus on substantial increases in R,D&D funding, the encouragement of European industry, market stimulation by subsidies, tax reliefs and premium payments, the setting up of regulations favouring Renewable energy penetration in utilities, municipal systems, the building sector, encouragement to include biofuel in non-food agricultural production schemes, and so on, as outlined in this Report.

Finally, Renewables must be viewed as 'alternative' only to traditional fossil fuel sources. They are, in fact, complementary to each other - and can be used effectively alone or in combinations of two or more (wind and biomass, for example). Thus all Renewable options should be pursued in tandem.

Bibliography / References

AFME (French Agency for Energy Management), Energy Management: A Strategy for the Future

Bettini V, Committee on Energy, Research and Technology, European Parliament Draft Report on the Promotion of Renewable Forms of Energy by the Foundation of a European Association for the Promotion of Renewable Energy, 3 June 1992, A3-0000/92 DOC_EN\PR\207488 PE 201.134 Or. IT

Bettini V, European Parliament, Working Document on the Promotion of Renewable Forms of Energy by the Foundation of a European Association for the Promotion of Renewable Energy, 6 January, 1992, (BS-1686/90), DOC_EN\DT\118314 PE 154.239/REV Or. IT

Beurskens J, Wind Energy systems: Resources, environmental aspects and costs, Electricity and the Environment - Background Papers for a Senior Expert Symposium held in Helsinki, 13-17 May, 1991, IAEA-TECDOC-624, IAEA, September 1991

Beurskens J, ECN, A Vision about Future Wind Energy Developments, Billund 16-10-1992

Cavanagh J, Electricity generation from Biomass, Electricity and the Environment - Background Papers for a Senior Expert Symposium held in Helsinki, 13-17 May, 1991, IAEA-TECDOC-624, IAEA, September 1991

CEC, Proceedings of the CEC Euroforum - New Energies International Congress, held at Saarbrucken, Germany, 24-28 October 1988, Volumes 1-3, EUR 11884, Stephens, 1988

CEC, Proposal for a Council Decision concerning the promotion of energy efficiency in the Community, COM(90) 365 final, Brussels, 13 November, 1990

CEC Directorate General for Energy, Energy in Europe Magazine, 1990 to date

CEC DGXII, Solar Europe - Photovoltaic, Building, Biomass, Wind - Newsletter, October 1992, Systemes Solaires, 1992

CEC, DGXII, Research and Development Programmes on Solar Energy Applications to Buildings - Research Digest Numbers 1-3

Cross B (ed.), European Directory of Renewable Energy Suppliers and Services, 1991 and 1992, James and James Science Publishers, 1991 and 1992

Danish Energy Agency, Ministry of Energy, Renewable Energy Technologies - Research, Technological Development, Enterprises 1992, 1992

Eichelbronner M, The Potential of Renewable Energyies in Germany - A Review of 1990 Studies and Scenarios, 1991

Entech Newsletter, 1988 to date

EOLAS, Renewable Energy Sources - Contribution to Ireland's Eneryg Supply in 1989: A Report Prepared for SOEC by EOLAS Energy Programme, October 1991

ETSU/NORWEB, Prospects for Renewable Energy in The Norweb Area, Crown, 1989

Eurostat, Eurostat D1 - Statistics on Renewable Energies in the European Community, 1992

Eurostat, Rapid Reports - Energy and Industry, 1990 to date

Eurostat, Yearbooks, 1990 to date

Eurostat, Monthly Statistics, 1991 to date

114

Eurostat, <u>Energy Balance Sheets</u>, 1991 to date

EWEA (European Wind Energy Agency), <u>Wind Energy in Europe</u>, October 1991

FhG ISE (Frauhhofer-Institut fur Solare Energiesysteme), <u>Transparent Insulation Technology for Solar Energy Conversion (Second Edition)</u>, 1991

Giolitti A, Jez S, Pisani A, Tiezzi E, Ulgiati S, <u>Power from biomass: land, energy and environmental constraints</u>, 1991 in Grassi G (ed.), Proceedings of the First European Forum on Electricity Production from Biomass and Solid Wastes by Advanced Technologies, 27-29 November, 1991, EUR 14689 EN, 1992

Goulding JR, Lewis JO, Steemers TC (eds.), <u>Energy Conscious Design: A Primer for Architects</u>, CEC Publication No. EUR 13445, Batsford for the CEC, 1992

Grassi G (ed.), <u>Proceedings of the First European Forum on Electricity Production from Biomass and Solid Wastes by Advanced Technologies</u>, 27-29 November, 1991, EUR 14689 EN, 1992

Grassi G, CEC, <u>Electricity Production from Biomass in the EC - Technological Options and Economic Perspectives</u>, dated February 1992, in Grassi G (ed.), Proceedings of the First European Forum on Electricity Production from Biomass and Solid Wastes by Advanced Technologies, 27-29 November, 1991, EUR 14689 EN, 1992

Grassi G, Bridgwater AV (eds.), <u>Biomass for energy, industry and the environment - a strategy for the future</u>, CEC Publication No. EUR 12897 EN, Edizioni Esagono, Italia, 1990

Grassi G, Trebbi G, Pike DC (eds.), <u>Electricity from Biomass</u>, CPL, 1992

Hall DO, Rosillo-Calle F, Biomass, Bioenergy and Agriculture in Europe, 1990, in Grassi G, Trebbi G, Pike DC (eds.), Electricity from Biomass, CPL, 1992

Hill R, The Potential Generating Capacity of PV-Clad Buildings in the UK, Contract No. E/5A/1365/2716 AEA Technology, May 1992

Hohmeyer O H, Macroeconomic View of Renewable Energy Sources, SunWorld, Volume 13, No. 2, 1989

Hohmeyer O H, Social Costs of Energy, Springer, 1988

IAEA, Senior Expert Symposium on Electricity and the Environment: Key Issues Papers, 13-17 May, 1991, Helsinki, Finland, 1991

Imrie SJ, The Environmental Implications of Renewable Energy Technology - Summary Report, 10 September 1992, Luxembourg PE 158.454, European Parliament, Directorate General for Research (Directorate B) - The STOA Programme, 1992

ISES, Solar Energy Solutions for an Environmentally Sustainable World - Recommendations of the United Nations Conference on Environment and Development, 1992

Jensen, SA, Implementation of Off-Shore Wind Power Plant, in Proceedings of the 1990 European Community Wind Energy Conference, 10-14 September, 1990, Spain, CEC Publication No. EUR 13251, Stephens and Associates, 1990

Johansson TB, Kelly H, Reddy AKN, Williams RH (eds.), Renewables for Fuels and Electricity: Renewable Fuels and Electricity for a Growing World Economy - Defining and Achieving the Potential.

Krause F, Bach W, Koomey J (eds.), IPSEP Final Report: Energy Policy in the Greenhouse -Vol.1, September 1989

Lipman N, Twidell J, Paynter R, Foster J (eds.), Renewable Energy - A Clean Technology: a Review and Strategy for SERC, 1992

Long G, Solar Aided District Heating Systems in the UK - An Appraisal, ETSU S 1190, 1992

Luque A, Sala G, Palz W, Santos GD, Helm P, Proceedings of the 10th. EC Photovoltaic Solar Energy Conference, Kluwer Academic Publishers, 1991

Madsen PH, Lundsager P (eds.), EWEA (European Wind Energy Agency), Proceedings of the European Wind Energy Association Special Topic Conference 1992: The Potential of Wind Farms, 8-11 September, 1992, Denmark, 1992

Mertens R, Nijs J, van Overstraeten R, Palz W, Technical Goals and Financial Means for PV Development, Proceedings of the 11th. EC Photovoltaic Solar Energy Conference, 1992

OECD, Proceedings of the Executive Conference on Photovoltaic Systems for Electric Utility Applications: Opportunities Critical Issues and Development Perspectives, 2-5 December, 1990, Taormina, Italy, OECD, Paris, 1992

Palz W (ed.), European Solar Radiation Atlas, Volumes I (Second Edition, 1984), and II, 1984, Verlag TUV Rheinland, Germany

Palz W, Renewable Energy in Europe, Int. J. Solar Energy, 1990, Vol.9, pp. 109-125

Palz W (ed.), Proceedings of the 1990 European Community Wind Energy Conference, 10-14 September, 1990, Spain, CEC Publication No. EUR 13251, Stephens and Associates, 1990

Palz W, CEC, Towards the Implementation of Renewable Energies - A European Perspective, Int. J. Solar Energy, 1991, Vol. 10, pp. 229-241

Palz W, Towards the Implementation of Renewable Energies - A European Perspective, in Scheer H (Editor-in-Chief), EUROSOLAR / UNESCO / CEC, The Yearbook of Renewable Energies, 1992 - Actions, Events, Initiatives, Ponte Press Bochum, 1992

Palz W, Schmid J, Technical Note: Electricity Production Costs from Photovoltaic Systems at several selected sites within the European Community, Int. J. Solar Energy, 1990, Vol. 8, pp.227-231

Palz W, Zibetta H, CEC, Energy Pay-Back Time of Photovoltaic Modules, Int. J. Solar Enregy, 1991, Vol. 10, pp. 211-216

Palz W, Zibetta H, CEC, Energy Pay-Back Time of Photovoltaic Modules, in Scheer H (Editor-in-Chief), EUROSOLAR / UNESCO / CEC, The Yearbook of Renewable Energies, 1992 - Actions, Events, Initiatives, Ponte Press Bochum, 1992

Piel J, Scientific American Special Issue: Energy for Planet Earth, Volume 263, No. 3 September 1990

The Renewable Energy Advisory Group - DTI, Report to the President of the Board of Trade, November, 1992, Energy Paper No. 60, Department for Enterprise, UK.

Saarbrucken, Usage du bois comme source d'energie, 1988, in Grassi G, Trebbi G, Pike DC (eds.), Electricity from Biomass, CPL, 1992

Scheer H (Editor-in-Chief), EUROSOLAR / UNESCO / CEC, The Yearbook of Renewable Energies, 1992 - Actions, Events, Initiatives, Ponte Press Bochum, 1992

Schlomann B, Environmental and Societal Costs, in Scheer H (Editor-in-Chief), EUROSOLAR / UNESCO / CEC, The Yearbook of Renewable Energies, 1992 - Actions, Events, Initiatives, Ponte Press Bochum, 1992

Schmid J, Klein HP, CEC, <u>Performance of European Wind Turbines - A Statistical Evaluation from the European Wind Turbine Database EUROWIN</u>, EUR 13929 EN, Elsevier Applied Science, 1991

Shock RAW, Palz W, CEC, <u>Wind Energy Generation Costs in Europe</u>, Int. J. Solar Energy, 1990, Vol. 9, pp. 57-63

<u>Solar Update Newsletter</u>, 1988 to date

Steemers TC, ECD Partnership, CEC, DG XII, <u>Passive Solar Energy as a Fuel: a Study of the Current and Future Use of Passive Solar Energy in Buildings in the European Community</u>, CEC Publication No. EUR 13094, 1990

Steemers TC, ECD Partnership, CEC, DG XII, <u>Passive Solar Energy as a Fuel: a Study of the Current and Future Use of Passive Solar Energy in Buildings in the European Community: Report on Background Material and Research</u>, CEC Publication No. EUR 13095, 1990

Steemers, ECD Partnership, <u>Solar Architecture in Europe: Design, Performance and Evaluation</u>, CEC Publication No. EUR 12738 EN, Prism for the ECD Partnership and the CEC, 1991

<u>Sun at Work in Europe Journal</u>, 1990 to date

Szeless A, <u>Solar Photovoltaic Electric Technology</u>, Electricity and the Environment - Background Papers for a Senior Expert Symposium held in Helsinki, 13-17 May, 1991, IAEA-TECDOC-624, IAEA, September 1991

Trench B, <u>Too Close to the Wind?</u>, Aer Lingus CARA Magazine, September/October 1992

Tributsch H, <u>Living with the Sun</u>, 1992, in Scheer H (Editor-in-Chief), EUROSOLAR / UNESCO / CEC, The Yearbook of Renewable Energies, 1992 - Actions, Events, Initiatives, Ponte Press Bochum, 1992

Troen I and Petersen EL, European Wind Atlas, Published for the CEC by Riso National Laboratory

Turrent D, Baker N, Steemers TC, Palz W, ECD/CEC. Solar Thermal Energy in Europe - An Assessment Study, Solar Energy R&D in the European Community Series A Volume 3 Solar Energy Applications to Dwellings, EUR 8473, Reidel for the CEC, 1983

UNSEGED, A Comprehensive Analytical Study on Renewable Sources of Energy, in Scheer H (Editor-in-Chief), EUROSOLAR / UNESCO / CEC, The Yearbook of Renewable Energies, 1992 - Actions, Events, Initiatives, Ponte Press Bochum, 1992

U.S. Department of Energy, The Potential of Renewable Energy: An Interlaboratory White Paper SERI/TP-260-3674 DE90000322 Prepared for the Office of Policy, Planning and Analysis, U.S. Department of Energy, March 1990

Van Overstraeten RJ, Proposals for Increased Use of Photovoltaics, Freiburg, September 1989

The Wright Commission Report

Zervos A, Palz W, Technological and Economic Development Outlook for Renewable Energy Sources for Electricty Generation, SM-323/21, Reprint from Electricity and the Environment - Proceedings of a Senior Expert Symposium on Electricity and the Environment, IAEA, 1991

Zweibel K, Harnessing Solar Power - The Photovoltaics Challenge, Plenum Press, U.S.A., 1990

List of Appendices

...Appendix 1: Choice of Renewable Energy Options

CHOICE OF RENEWABLE ENERGY OPTIONS

In limiting the choice of Renewable Energy options to the four that were chosen, the authors opted for those which in the medium to short term they felt were technologically feasible and commercially promising - as opposed to a detailed consideration of all Renewable Energy sources currently on offer. Different Renewable Energy technologies are at different stages of development. Some, such as both large and small scale hydro, are close to or have achieved technical maturity and need little further research and development.

There are other Renewable technologies which research and development activity to date has shown to be economically and technically far less promising than those which have been chosen. The use of geothermal acquifers is a promising technology - however, geothermal hot dry rock is evidently far more difficult to tap than was originally expected. Similarly, large off-shore wave energy devices, though conceptually appealing and, theoretically at any rate, with great energy potential, seem, for the moment at least, to have floundered on the rock of actual technological realization.

Solar thermal power also represents a promising technology - although within Europe, it is mainly applicable to the southern Mediterranean regions. Medium and high temperature thermal use of solar energy may be a viable energy source in the future for two kinds of application: (a) production of electric energy; (b) process heat for the production of fuels and chemicals. Medium concentration devices (e.g. parabolic trough concentrators) have typical temperatures of 300-500° C while high concentrating devices (parabolic dish concentrators, solar tower arrangements) have temperatures of up to and more than 1000° C . For both of these types, a number of test facilities exist in southern Europe and medium concentrating electric power plants are stated to be close to commercialization in the US. A number of technical problems remain to be overcome, however, before full commercialization is possible in Europe and although the long term prospects for this particular technology seems bright, with predicted electricity costs of 0.1 ECU/kWh, its lack of potential in the medium to short term as well as its applicability only to the very southern parts of Europe, nevertheless precluded its consideration in the present study.

...Appendix 2: EUR-12 Primary Energy Production/Consumption 1991

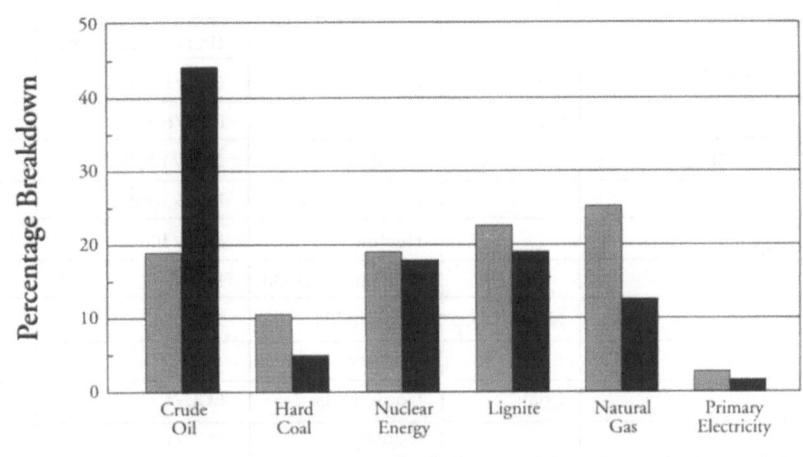

Total Production: 623.9 MTOE
Total Consumption: 1201.9 MTOE

...Appendix 3: Energy Import and Final Energy Consumer Prices 1991

			Unit of Measurement	Prices 2Q91
IMPORT PRICES	Crude Oil	(cif)	ECU/barrel	17.94
	Steam Coal		ECU/tce	51.10
FINAL CONSUMER PRICE	Oil Products	Gasoline	ECU/1000lt	744.00
		Diesel	ECU/1000lt	480.00
		Heating Oil	ECU/1000lt	323.00
		Residual Fuel Oil	ECU/t	108.00
	Natural Gas	Households	1984=100	104.90
		Industry	1984=100	70.60
	Coal	Households	ECU/t	206.50
		Industry	ECU/t	93.10
	Electricity	Households	ECU/100KWh	12.07
		Industry	ECU/100KWh	6.50

TOTAL EUR-12
CO_2 SO_2 AND NO_x
EMISSION LEVELS
1991

	EUR-12 (old)	EUR-12 (new)
CO_2	2765.9 mtons	3042.3 mtons
SO_2	12,117,000 tons	15,555,000 tons
NO_x	11,520,000 tons	12,134,000 tons

Source: Energy in Europe Magazine

...Appendix 5: Energy in Europe 1990-2000:
Four Possible Energy Scenarios / Total Primary Energy Requirement

	Principal Features
Scenario 1	- steady economic growth - just gradual policy development - technological development - improved efficiencies - market forces driving system within existing frameworks
Scenario 2	- high economic growth without appropriate policy measures and based only on market mechanisms - supply capacities under pressure - high levels of polluting emissions
Scenario 3	- mastery of energy consumption - more efficient means of production - high economic growth with - strict environmental standards
Scenario 4	- moderate economic growth but with stricter environmental standards than Scenario 1 - mastery of energy consumption and - more efficient means of production as in Scenario 3

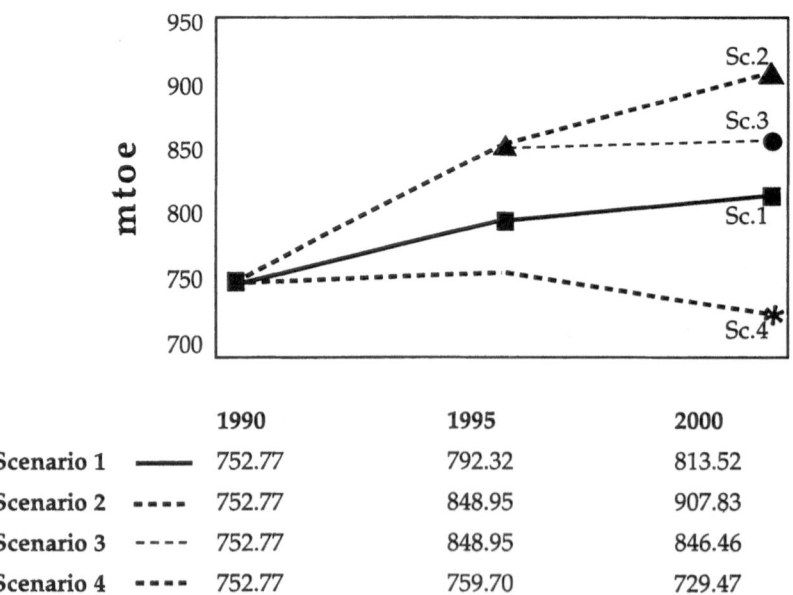

		1990	1995	2000
Scenario 1	———	752.77	792.32	813.52
Scenario 2	- - - -	752.77	848.95	907.83
Scenario 3	- - - -	752.77	848.95	846.46
Scenario 4	- - - -	752.77	759.70	729.47